The POSCO Strategy

The POSCO Strategy

A Blueprint for World Steel's Future

William T. Hogan, S.J.

LEXINGTON BOOKS
Lanham • Boulder • New York • Oxford

LEXINGTON BOOKS

Published in the United States of America
by Lexington Books
4720 Boston Way, Lanham, Maryland 20706

12 Hid's Copse Road
Cumnor Hill, Oxford OX2 9JJ, England

British Library Cataloguing in Publication Information Available

Library of Congress Cataloging-in-Publication Data

Hogan, William Thomas, 1919–
The POSCO strategy : a blueprint for world steel's future / William T. Hogan.
p. cm.
Includes bibliographical references.
ISBN: 978-0-7391-0301-2
1. P°'hang Chonghap Chech°'l Chusik Hoesa. 2. Steel industry and trade—Korea (South). I. Title.

HD9526.K64 P638 2001
338.4'7669142'095195—dc21 2001038567

Printed in the United States of America

∞™ The paper used in this publication meets the minimum requirements of American National Standard for Information Sciences—Permanence of Paper for Printed Library Materials, ANSI/NISO Z39.48-1992.

This book is dedicated to Sang-Boo Yoo, Chairman of the Board and Chief Executive Officer of POSCO. He has guided the destinies of POSCO for the past four years in an eminently successful manner, particularly when the steel industry in many parts of the world was experiencing difficulty. He has further laid out plans which should continue POSCO's growth as a leading steelmaker into the foreseeable future. The operation of POSCO has made a distinct contribution to the world's steel industry and has also been a dominant force in the industrialization of Korea.

Contents

Acknowledgments

Over the years, I have had extensive conversations with the chairmen of POSCO, including Tae-Joon Park, Kyung-Ro Hwang, Myung-Sik Chung, Mahn-Je Kim, and Sang-Boo Yoo, all of whom have my sincerest thanks for so generously sharing their visions and strategies for their company's success. Likewise, I would like to thank Frank T. Koelble, the Associate Director of the Industrial Economics Research Institute, for valuable assistance in preparing much of the research for this book. I am also indebted to the POSCO organization in Korea for a very significant amount of detailed information, as well as to the POSCO people in the United States, including Moon Q. Kim. I also want to acknowledge Lisa Smith of Ward's Communications for information on the Korean automobile industry. A considerable amount of help also was provided by Michael D. Tusiani of Poten & Partners, Inc. and Mogens Petersen of his organization, who provided significant information on the shipbuilding industry. Likewise, Michael T. McGuire of Voest-Alpine Industries, Inc. contributed much necessary information. Thanks also is due to Mary B. Murtagh, who worked on the manuscript from the start, and to Peter B. Mulloney, William S. Kirk, and Glennon C. Kidd, who so generously shared their knowledge of the steel industry.

Introduction

The Republic of Korea is an outstanding example of a once nonindustrialized country that has developed a large, modern steel industry in less than 25 years. In 1966, the nation's steel output was a minuscule 185 thousand tonnes, produced in small, obsolete furnaces. In 1970, production, entirely from small, nonintegrated plants, many of which used electric furnaces, amounted to only 481 thousand tonnes. Then, in one of the truly remarkable modern-day industrial achievements, the country by 1992 had become the world's seventh-largest steel producer with an output of 28.1 million tonnes. Its production in that year surpassed that of such industrialized countries as Italy, France, the United Kingdom, Canada, Spain, Australia, and South Africa. This exceptional growth during a difficult period of retrenchment for world steel at large came from two new integrated plants and a number of new electric-furnace operations.

By October 1992, Pohang Iron and Steel Co., Ltd. or POSCO, had succeeded in constructing two fully integrated and profitable steel plants, which in 1993 turned out 22.5 million tonnes or about 70 percent of the nation's raw-steel production. This made POSCO the second-largest steel-producing company in the world.

Korea's steel industry has been a catalyst for the growth of other domestic industries, including automotive, appliances, construction, distribution through service centers, shipbuilding, and containers, and has moved to a position where it ranks high in the industrialized world. The country is obviously no longer a Third World nation, since many of its industries have attained a high ranking among those in industrialized nations, and POSCO now has become the largest steel company in the world.

The Korean experience, particularly that of POSCO's growth and development, represents an encouraging case study. POSCO's approach provides a strategy or blueprint for other steelmakers that demonstrates just how their industry still can prosper and why it remains indispensable to other industries that, together with steel, drive economic growth and progress around the world. POSCO's success underscores the potential for positive future opportunities in utilizing the latest iron and steel technologies to better satisfy the needs of manufacturers and other steel consumers. By proving that effectively executed steel ventures can generate continuously positive returns, even in times of recurring difficulty, POSCO has given world steel, as well as its customers, suppliers, investors, and government policymakers a convincing reason to be more positive in viewing the industry's future.

How does a company originating in a nonindustrialized country move from virtually last to first in the world in so short a period of time? How does it manage to earn a profit year in and year out, even at times when a majority of its counterparts in other countries are mired in losses and financial difficulty? The answers to these questions are to be found in what can be termed "The POSCO Strategy," which is the title and subject of this book. The subtitle, "A Blueprint for World Steel's Future," recognizes the significant benefit that other steelmakers can derive by studying key aspects of the strategy, which are presented and examined in the chapters that follow.

As used throughout the text, Korea refers to the Republic of Korea or South Korea, and unless otherwise indicated, the term "tonne" is used to designate a metric ton of 1,000 kilograms.

1

A Shift in Strategies

The POSCO strategy actually had its genesis years before POSCO's founding in 1968. By then, Korea's leaders already had made a number of unsuccessful attempts to construct a modern steelworks in their steel-deprived nation, an objective that had been made a central component of the government's 1962 and 1967 five-year economic plans. These early efforts not only demonstrated Korea's resolve to employ steel as a means of accelerating the industrial development of the nation's economy, but they also gave rise to deeply held views among the Koreans involved about how this could best be accomplished, views that eventually would crystallize into a strategy that was to make POSCO the world steel industry's most successful company.

But first there was another steel strategy that came to be abandoned, one largely devised by a consortium of international steel interests known as Korea International Steel Associates or KISA. Korea's President Chung-Hee Park had asked Koppers Company, Inc. of the United States to organize the consortium in conjunction with advancing the second five-year economic plan for 1967–71, and KISA was formed toward the end of 1966, with a final agreement signed in March, 1967, whereby the consortium was to "design, engineer, supply, construct, and commission an integrated iron and steelmaking facility for the Republic of Korea."[1]

KISA's member organizations were a who's who of the world's leading steel-mill builders and engineering firms, including three from the United States, namely Koppers, Blaw Knox Company, and Westinghouse Electric International Company; two from Germany, DEMAG Aktiengesellschaft and Siemens Aktiengesellschaft; and one each from England, Italy, and France, namely the Wellman Engineering Corporation, Ltd., Societa Italiana Impianti, S.p.a., and ENSID, respectively.

It was understandable and not insignificant that KISA's members were drawn exclusively from industrialized countries in which most of the leading steel technologies of the mid-1960s resided. The same countries housed many of the world's large, integrated steel plants, which had capacities sized in the millions of tonnes to attain required scale economies and effective production costs. Comparable plants were uncommon in developing countries, which undoubtedly colored the consortium's shared concept of just what a Korean integrated steel plant should be.

Even before KISA was formed, the International Bank for Reconstruction and Development, more commonly known as the World Bank, was already on record regarding the proposed plant's size, having concluded in a May 1966 technical report to the Korean government that it would be feasible to construct a facility capable of producing 500 thousand tonnes of steel annually.[2] Such a plant would, however, leave national steel output nearly a million tonnes short of the goal established in Korea's second five-year economic plan.

In retrospect, a strategy whereby KISA was to build a steel plant "for" Korea probably preordained the insurmountable problems that such a project was to encounter, particularly in regard to the consortium's recommendations for the proposed plant's capacity and product capabilities. From the project's early stages, the views of the Korean government and those of the nation's economic-policy and business leaders generally coalesced around a steel plant of much more ambitious proportions than that prescribed by the consortium and various international lending agencies. The disparity had much of its basis in differing national and business self-interests, which in time were to see KISA's attempted steel project unsuccessfully concluded. However, the experience served to harden the determination of those on the Korean side to build the integrated steelworks they wanted, and not the one others wanted them to build.

Appointed by President Chung-Hee Park in September 1967 to represent Korea's interests in the KISA project was General Tae-Joon Park, whose eventual role in building the nation's steel industry was to make him the Carnegie, Thyssen, or Inayama of his day. In 1967, General Park was president of Korea Tungsten Mining Company, a onetime money-losing firm he had succeeded in turning around to become Korea's number one earner of foreign exchange. Even before the KISA agreement was signed, he had been asked by President Park to begin preparations to construct a steel plant. Now he was named Chairman of the Promotion Committee for the Integrated Steel Mill and was later assigned the tasks of securing investment capital for the mill and heading up its construction.

An important consideration in General Park's appointment was the close relationship he had established with steelmakers in Japan. This would afford access to their know-how, which had been kept from KISA by the Japanese government's refusal to permit Japanese firms to join the consortium. As it turned out, General Park's contacts in Japan were to play a far more significant role than was anticipated, one that would become pivotal to the funding and development of POSCO, the company that in time would put the Korean steel industry on the map.

THE KISA STRATEGY

The integrated plant that KISA intended to build for Korea was described in summary form by the following terms of reference: The plant was to be installed in two stages, the first including all the plant and equipment components required to produce 500 thousand tonnes of raw steel per year, with the second stage planned for expanding the plant's facilities to produce 1.0 million tonnes of raw steel per year. Construction schedules anticipated commissioning the first stage completely during 1971 and the second stage in 1976, "depending upon the economic growth of the Republic."[3]

Shaping this strategy in terms of both plant capacity and scheduling was the influence of the World Bank and other international lending agencies that were being counted on to provide a major portion of the project's funding needs of more than $135 million, not counting additional infrastructure investments expected to be underwritten mainly by the Korean government. The World Bank had circumscribed the proposed plant's "feasible" capacity at 500 thousand tonnes primarily by evaluating various studies of the Korean steel market made during 1965–66. On average, the studies placed Korea's consumption of finished products in 1965 at 328 thousand tonnes, and their projections for 1971 varied from less than

500 thousand tonnes to 903 thousand tonnes. By contrast, Korea's second five-year economic plan placed increased emphasis on heavy industry and thereby projected significantly greater steel requirements.[4]

Under the second five-year plan, Korea's gross national product (GNP) during 1967–71 was programmed to grow at a 7 percent annual rate, conservatively established to be somewhat less than its actual growth had been under the first five-year plan. However, the manufacturing sector of the economy was expected to increase its share of total GNP from 22 percent in 1965 to 27 percent in 1971, and to continue an upward trend to 32 percent in 1981. Further, the ratio of heavy to light industry, which was 28 percent to 72 percent in 1965, was scheduled to become 34 percent to 66 percent in 1971.

According to the second five-year plan, the metal and machine industry was to undergo an eightfold expansion, "supported by 1.6 million tonnes of iron and steel production."[5] The plan called for accelerated and diversified industrialization, with emphasis on such industries as chemicals, machinery, and iron and steel. These manufacturing activities were considered essential to building the foundation for heavy industry and, in turn, to achieving balanced industrial growth in the future.

The wide disparity between the second five-year plan's call for 1.6 million tonnes of iron and steel output in 1971 and the proposed new steel plant's 500 thousand tonnes of capacity that was recommended by the World Bank and KISA entailed more than a disagreement over potential market growth. There were also problems posed by a small integrated plant's lack of the economies of scale, a shortcoming that KISA addressed as follows: "Historically, an integrated steel plant, having an annual capacity of 500 thousand tonnes of crude steel, producing semi-finished products is not ordinarily a viable undertaking. Such might well have been the case in Korea, but this has been avoided by phasing construction toward production of one million tonnes and the early importation of slabs, thereby substantially increasing the rolling mill output to a profitable degree. After an examination of possible outlets for steel products, the Project Development Plan was based on facilities to produce continually cast billets for re-rollers, finished plates, skelp for pipe and hot-rolled strip. This program offers a balanced operation between steelmaking and rolling facilities."[6]

In this manner, the KISA strategy attempted to finesse the conflict between the capacity limitations imposed by the World Bank and the technical realities of integrated iron and steel production. KISA relied on the proposed plant's uncertain second phase, which had been made contingent upon "economic growth in the Republic," to explain away the inherent problems of building an undersized integrated plant in the first phase. Moreover, while the proposed plant was to produce low-value-added semifinished billets for others, it would have to import semifinished slabs to give its own rolling operations a chance at profitability.

Producing billets for Korea's re-rollers was seen as providing one justification for building an integrated plant, since this was expected to improve the product quality of Korean steel producers then using electric furnaces to melt questionable scrap as a feed material for their operations.

KISA'S PROPOSED PLANT

The process flow of raw materials and products for KISA's proposed 500-thousand-tonne integrated plant is shown in the stage-one diagram in Figure 1.1. Starting at the unloading dock, cranes and equipment were to be provided to handle iron ore, coke, and limestone at an unloading rate of 1,300 tonnes per hour. In stage one, the proposed plant was to operate without coke ovens, so that all of the coke required for ironmaking would have to be imported. A gantry load-out crane with a lifting capacity of 22 tonnes was to be used for loading billets, plates, and strip coils produced at the plant for ship transport. Materials handling facilities were to be installed to transfer the raw materials from the dock after unloading to stockpiles, prior to their movement to the blast furnace and calcining plant. Both rail cars and trucks were to be used for this purpose.

The plant's ironmaking section was to have a single 7.62-meter (25-foot) hearth diameter blast furnace for producing the plant's stage-one hot metal requirements of 520 thousand annual tonnes. Most of this output was to be moved to the steelmaking shop in torpedo cars, and a portion was to be processed in a pig-casting machine. The steelmaking shop was to consist of two 60-tonne, top-blown, basic-oxygen furnaces (BOF), ingot-teeming facilities, a six-strand continuous caster for billets, and a lime calcining plant. Based on operating one of the two BOFs, the shop was to produce 500 thousand tonnes of liquid steel annually, pour some

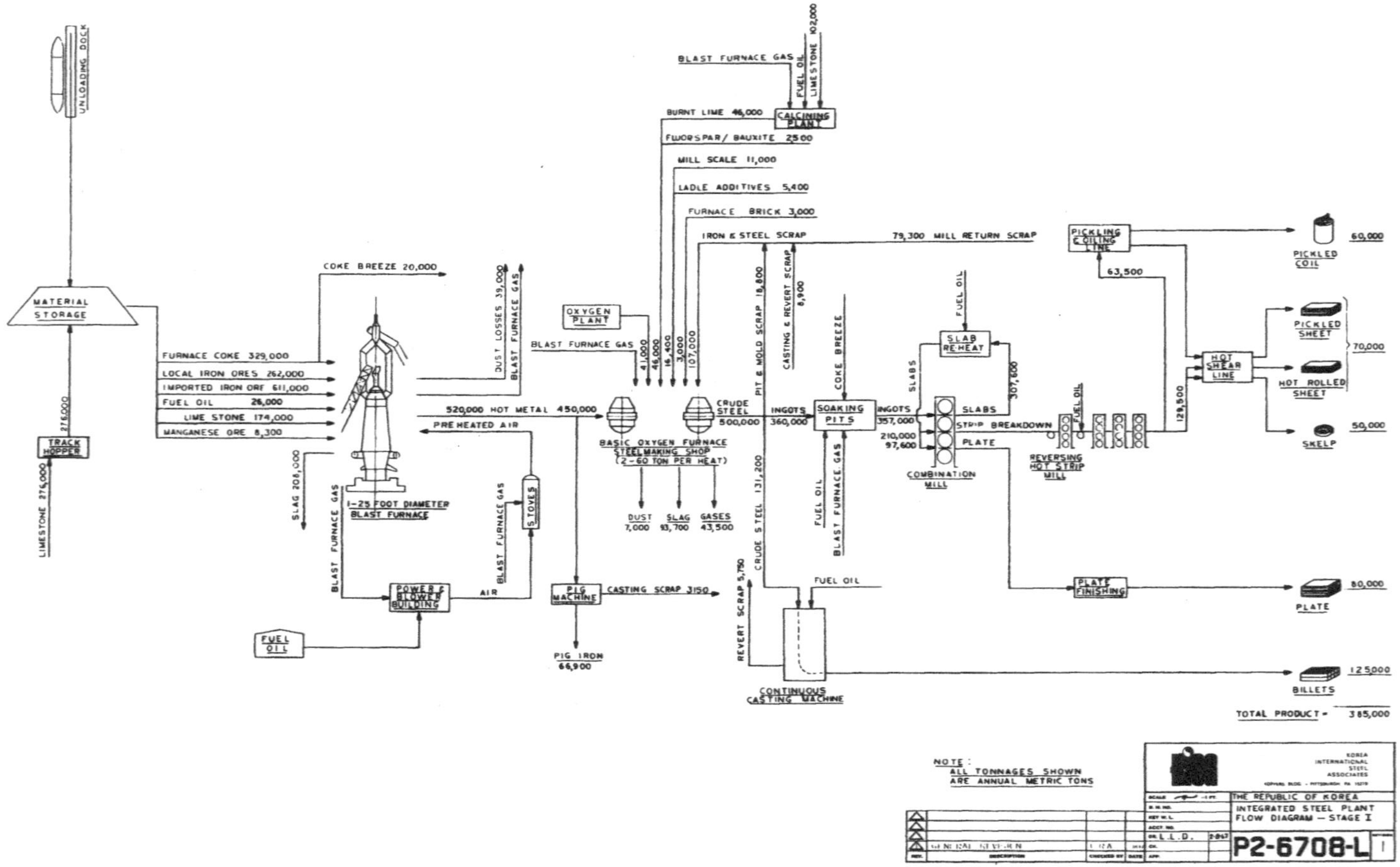

Figure 1.1 Stage One of the Proposed KISA Plant

358 thousand tonnes of ingots, and continuously cast 125 thousand tonnes of billets.

The stage-one plant's rolling-mill section was to consist of six ingot-soaking pits and a single three-zone, continuous slab reheating furnace rated at 100 tonnes per hour; a two-way combination slabbing and plate mill, convertible to either a two-high, 46-inch, high lift breakdown mill for rolling ingots into slabs, or alternatively to a four-high, 96-inch plate mill; a four-high, 56-inch reversing hot-strip mill to produce hot-rolled coils; and finishing equipment for plates and hot strip, including a 56-inch continuous pickling and oiling line with a side trimmer. Beyond the plant's major production sections, general facilities were to be provided, including all auxiliary, service, and utility equipment and connections.

The proposed plant's final stage-one output was to total 385 thousand annual tonnes of steel products, including the aforementioned 125 thousand tonnes of billets, 80 thousand tonnes of plates, 70 thousand tonnes of hot-rolled and pickled sheets, 60 thousand tonnes of pickled coils, and 50 thousand tonnes of skelp for pipemaking. At Korea's behest, proposed stage-one product production was later increased to some 460 thousand tonnes by raising the plant's initial steelmaking capacity to 600 thousand tonnes.

KISA's original review of potential seacoast sites for the proposed plant had reduced its recommended alternatives to two locations, one at Sam Cheonpo on the Korea Strait and the other at Ulsan in the northern coastal region. However, other Korean and overseas survey teams were also investigating alternative sites, and in June, 1967, the Korean government announced that Pohang had been selected as the location for building the plant. Situated on Korea's east coast on the Japan Sea at Yongil Bay, the Pohang site sloped gradually from the shore to deep water, and although unprotected and without a natural harbor, the water's depth gave the site maximum potential for constructing a port that could accommodate the largest bulk cargo carriers then in use. This selection of a site affording unlimited access to water transport was one sign that Korea's objectives for its steel industry did not share the capacity constraints common to KISA's strategy.

In November 1967, the Korean government coordinated efforts to secure essential outside financing for the steel project and formed the Steel Mill Construction Committee to work with Korea Tungsten in setting the project in motion at the Pohang site. The following spring, as site preparation work at Pohang progressed, a new steel corporation, Pohang Iron and Steel Company, Ltd., or POSCO, came into being, with General Park elected its president. Meanwhile, government representatives continued to negotiate with the World Bank, the Export-Import Bank, and others in attempting to arrange project financing. As the months wore on, the lack of progress in these negotiations became an unavoidable sign that the KISA project was in serious trouble.

SEEKING NEW SUPPORT

It soon was the start of 1969, and, despite continuing, intense negotiations, firm financial commitments in support of KISA's proposed plant still were proving elusive. The World Bank had started to express concern over the risks of funding an integrated steel plant in a developing country with limited economic resources, a weak balance of payments, and a limited ability to repay the required loans. In March 1969, this concern was cited by the World Bank in reversing its May 1966 conclusion that a Korean plant with 500 thousand tonnes of capacity was feasible. Instead, the World Bank joined the Export-Import Bank in encouraging Korea to abandon its efforts to build a steel plant in favor of first developing a machinery industry. The machinery alternative was portrayed as more beneficial to Korea in terms of utilizing both technology and labor, but also was said to entail less risk to potential lenders.

Beyond the World Bank's eleventh-hour reversal on the issue of economic feasibility, the KISA project was also objected to by the U.S. Agency for International Development (AID), ironically because the limited capacity of the proposed plant precluded taking advantage of the economies of scale. The project thus became conflicted between the capacity constraints imposed by the World Bank's market researchers, on the one hand, and the much larger capacity needed to tap into scale economies on the other.

According to the World Bank's revised view, Korea could no longer afford to build a 500-thousand-tonne plant, much less a plant capable of producing millions of tonnes. However, even if the larger plant could be afforded, AID was concerned that adding so much capacity in Korea would be disruptive to the structure of world steel prices. Accordingly, AID recommended that Korea focus on installing either a smaller, electric-furnace-based steel plant or new rolling facilities unsupported by steelmaking, such as a continuous hot-strip mill that would be fed totally with imported slabs.

Along with the rejection by U.S. banking agencies, the British and German governments also removed their potential backing, bringing the KISA project to a practical end. Clearly, if Korea was to build an integrated steel plant, an alternative strategy would have to be adopted, starting with a search for another means of financial support. And this is where the Japanese contacts of POSCO's General Park were to demonstrate their true worth.

Throughout the spring and summer of 1969, General Park focused on seeking Japanese support for what was now the POSCO project. His main objective was to have Japan redirect a portion of its war-reparations payments to Korea into helping construct a steel plant at Pohang. Some $300 million in war reparations had been earmarked to foster Korea's agrarian development over a ten-year period starting in 1966. However, General Park proposed that some of these funds together with low-interest loans from Japan's Overseas Economic Cooperation Fund be provided to meet the POSCO project's foreign capital requirements of some $125 million. This would replace the funding denied by U.S. and European banking agencies and make Japan the sole overseas supporter of the project.

The case for the POSCO project was made by General Park in meetings with the chairmen of Japan's three largest integrated steel companies, Yawata, Fuji, and NKK. He argued that industrialization with steel at its base was the key to building a vigorous Korean economy, which was then limited in its potential by its dependence on steel imports. The three chairmen, having seen firsthand just how important the development of their own steel industry was to Japan's economy, eventually agreed to use their influence in convincing Japan's government to redirect the required war-reparations funding. Pending this approval, they agreed to have the Japan Iron and Steel Federation provide technology assistance and construction support to the project at Pohang.

By the fall of 1969, the funding plan had become a reality, and the POSCO strategy was about to unfold.

NOTES

1. Korea International Steel Associates (KISA), *Project Development Plan*, March 1967, p. I-1.

2. For a discussion of events surrounding the KISA project see J.J. Innace and A. Dress, *Igniting Steel* (New York: Global Village Press, 1992), p. 31ff.; and K.K. Seo, *The Steel King* (New York: Simon & Schuster, 1997), p. 107ff.

3. KISA, *op. cit.*, p. I-2.

4. KISA, *op. cit.*, pp. I-11ff. and II-6ff.

5. KISA, *op. cit.*, p. I-12.

6. KISA, *op. cit.*, p. I-3.

2

The Building Strategy

POSCO's incorporation on April 1, 1968, as a private company immediately distinguished the aspiring steelmaker from many of its counterparts in other countries, both developing and industrialized, where control of the steel industry was shifting into government hands.

The new company's organizational structure sought to realize the benefits of both corporate independence and government financial support. This was accomplished by incorporating the company under Korean commercial law as a private enterprise, but one that would be financed under special legislation providing for majority government ownership. This approach, devised by General Park, ensured the management autonomy that was to prove essential to the company's success.[1]

A "private" POSCO did not intend to forgo major government financial backing, which would have been impossible considering Korea's then early stage of economic development and the ambitious scope of the new company's plans for entering the steel business. Rather, the decision to adopt private status underscored POSCO's intention to operate at a profit, an objective at the core of a strategy for building a highly competitive steel enterprise that would be an effective catalyst for Korea's industrialization, as well as a means of achieving national self-sufficiency in steel.

The adoption of a profit motive by a company that would remain largely government owned for the succeeding 20 years was designed to avoid the problems faced by other government-owned steelmakers. The profit motive was to establish a clear delineation between ownership and control, with decisions and strategies for plant construction and operation residing in company management and not with government policymakers.

In POSCO's case, for example, employment at its new steel plants was to be consistent with the highest worker productivity and not, as was the case elsewhere, to meet government standards for job creation that too often turned steel mills into high-cost employment agencies. In POSCO's case, capital equipment and consumables for its new steel operations were to be purchased by the company from the most competitive worldwide sources and not from government-specified local suppliers that may or may not be capable of meeting price and quality standards. And POSCO would insist on exclusive control over capital management to ensure that government agencies would not redirect project funding for other purposes, as had become common elsewhere.

From an operating standpoint, government ownership in other countries had been accompanied by serious pitfalls, particularly when the world-steel market was on the down slope of its recurring cycle. At such times, government policies to sustain employment by keeping steel operations at a high level had resulted in excess production, sharp cuts in domestic and export prices, and severe losses that were absorbed in one way or another by the government. This was an outcome that both POSCO and the Korean government wanted to avoid. And so, from the outset, POSCO, although largely government owned, sought to operate as a private, profit-based company that was determined not to incur losses for any extended period. This would be done by maintaining a price structure that in the long run would cover the company's total costs of production, which, in turn, would be kept as low as possible. Notably, when adopted by POSCO, this time-honored precept was in the course of abandonment by an increasing number of steelmakers around the world.

The intended adherence to pricing discipline and the profit motive was a primary consideration in POSCO's winning Japanese support for its first integrated plant. Given the close geographic proximity of the Japanese and Korean markets and the concerns that had been expressed by other potential lenders over the risks that a large-scale Korean integrated steelworks would pose for the structure of steel prices in world markets, Japanese steelmakers and their government undoubtedly were reassured that POSCO had resolved not to follow the overproduction and price-cutting practices that had become all too common among government-owned steel companies.

POSCO was incorporated and embraced the profit motive less than a year after the massive nationalization of British Steel Corporation. It was a time when the government-owned share of world-steel capacity was in the midst of climbing from just over 20 percent in 1950 on the way to 60 percent in 1980, which was in marked contrast to the more recent wave of steel privatizations that intensified after the collapse of the Soviet Union.

For POSCO, the objective of earning a profit transcended mere political considerations, affording not only a means to avoid the mistakes made by fledgling steel producers in other developing countries, but also a way to instill the highest standards into its pursuit of becoming a competitive steelmaking company of significant proportions. Recognizing that profits would have to be achieved within a corporate culture accommodating government ownership, POSCO's earliest strategy for building the company and its steelmaking capabilities was to first secure commitments from the government vesting authority in the company for all personnel recruitment, all purchasing, and the management of all finance capital. This represented a significant departure from the government control over these key activities that often had become the norm for steel projects in other countries.

Beyond the broad early strategy of operating as a private, profit-driven company, POSCO employed a variety of more specific strategies as it progressed from incorporation and into the funding, construction, and operating stages of its development. A chronology of major events in POSCO's history is set forth in Table 2.1, and the strategies instrumental in building the company are made apparent by discussing just how the company's two fully integrated plants were constructed, the first at Pohang and the second at Kwangyang.

THE POHANG PLANT

On April 1, 1970, two years to the day after POSCO was incorporated, ground was broken at Pohang to begin constructing the major facilities of the company's first integrated steelworks. By then, nearly a year and a half had been spent arduously re-sculpting the terrain at the 2,200-acre

Table 2.1 Chronology of Major Events in POSCO's History

April 1968	Pohang Iron and Steel Co., Ltd. founded
April 1970	Commenced 1st-stage construction of Pohang Works
July 1973	Completed 1st stage of Pohang Works (1.03 million tpy crude steel)
October 1973	POSCO joins the International Iron and Steel Institute (IISI)
May 1976	Completed 2nd stage of Pohang Works (2.6 million tpy crude steel)
December 1978	Completed 3rd stage of Pohang Works (5.5 million tpy crude steel)
February 1981	Completed 4th stage of Pohang Works (8.5 million tpy crude steel)
May 1983	Completed 2nd phase of 4th stage of Pohang Works (9.1 million tpy crude steel)
March 1985	Commenced 1st-stage construction of Kwangyang Works
April 1986	USS/POSCO Industries joint venture formed
March 1987	Research Institute of Industrial Science & Technology founded
May 1987	Completed 1st stage of Kwangyang Works (11.8 million tpy crude steel)
June 1988	POSCO stock listed on the Korea Stock Exchange (KSE)
July 1988	Completed 2nd stage of Kwangyang Works (14.5 million tpy crude steel)
December 1990	Completed 3rd stage of Kwangyang Works (17.5 million tpy crude steel)
October 1992	Completed 4th stage of Kwangyang Works (20.8 million tpy crude steel)
December 1993	ISO 9002 certification awarded
October 1994	POSCO American Depositary Receipts (ADRs) listed on the New York Stock Exchange (NYSE)
October 1995	POSCO ADRs listed on the London Stock Exchange (LSE)
November 1995	Construction of COREX plant completed
October 1996	Construction of No. 1 Mini Mill at Kwangyang Works completed
August 1997	Construction of No. 4 Cold Rolling Mill at Kwangyang Works completed
April 1998	30th Anniversary of POSCO
March 1999	Construction of No. 5 Blast Furnace at Kwangyang Works completed (28.0 million tpy crude steel)
April 2000	Began operation of No. 5 Blast Furnace at Klwangyang Works
October 2000	Completion of Privatization

Source: Pohang Iron and Steel Company, Ltd.

site, relocating villages, laying a 9.5-mile rail spur, expanding area roadways, building bridges, and, despite extreme difficulties, creating a harbor for large bulk carriers, a raw-materials wharf, and a breakwater to tame Yongil Bay's angry tidal swells.

This marked the first time that a civil-engineering project of such massive proportions had been undertaken in Korea, so that a lack of domestic equipment and technology was a recurring obstacle, as was the initial inexperience of POSCO's own construction crews. From the outset, however, the company's building strategy required that its own workers fully participate in executing all phases of the plant's construction, which, in time, paid enormous dividends, not only from the workers' dedication, but just as importantly, from the know-how and skills imparted by the experience.

POSCO sought to learn everything possible from the technological cooperation extended by the Japanese steel industry, without which the Pohang plant's construction would not have been possible. This was accomplished by emphasizing worker training and education as an integral part of the construction process, and, thereafter, the plant's daily operating activities, a practice that has continued with great effectiveness to the present day.

The site at Pohang had been selected and prepared to accommodate the multi-stage construction of a large steelworks that ultimately would be able to produce many millions of tonnes. The First Stage of this steelworks had a designed steelmaking capacity of 1.03 million annual tonnes, and utilizing 24-hour workdays, its construction was completed ahead of schedule, within 38 months, on July 3, 1973. This gave POSCO more capacity than had been considered for the KISA project's Second Stage, which would not have come on stream until 1976 at the earliest.

The mid-1973 completion of the Pohang plant's First Stage established another hallmark of POSCO's building strategy, namely to always adhere to or better established construction schedules. Throughout the company's growth and development, now extending over 30 years, this objective always has been fulfilled, helping to keep construction costs as low as possible. Pohang's First Stage construction encompassed raw-materials unloading and handling facilities, one battery of coke ovens, a sinter plant, a single-blast furnace, a basic-oxygen steelmaking plant, a slabbing mill, billet mill, plate mill, and hot-strip mill. The capital expenditure required was held to a level unheard of by world standards at the time, approximately $296 million, representing a unit cost of only $287 per tonne of steelmaking capacity.

Contributing to the low cost of the First Stage were a number of other important tenets of POSCO's building strategy relating to the selection, purchase, and installation of iron and steel facilities. Although it had become common practice among steelmakers in developing countries to purchase used equipment as a means of reducing capital costs, POSCO concluded that the short-term benefit of doing so would soon be negated by the adverse impact on the company's long-term competitiveness. Consequently, the decision was made early on to purchase only new facilities and to insist on the best and latest technologies available. Instead of purchasing less-expensive used equipment, POSCO controlled its capital costs by obtaining multiple quotes in making facility purchases, by demanding excellence in execution from equipment suppliers from delivery through installation and startup, and by strictly negotiating finance terms and interest charges. The low capital cost of Pohang's First Stage proved convincingly that this approach could be successfully employed.

Immediately after the First Stage was dedicated, construction began on Pohang's Second Stage to increase steelmaking capacity to 2.6 million annual tonnes. Throughout the construction process, POSCO maintained full production at the facilities already in place in order to keep pace with rising domestic demand. The Second Stage was designed to add a second blast furnace and steelmaking shop and to scale up all the facilities of the First Stage except for the billet mill. In addition, a cold-rolling mill and galvanizing line were to be installed, and a continuous-casting plant was to be built, housing a single-strand slab caster capable of processing 400 thousand annual tonnes of slabs and a three-strand bloom caster rated at 270 thousand tonnes. The Second Stage was completed and dedicated within 30 months on May 31, 1976, and required a capital investment of $553 million, or some $352 per tonne.

POSCO had been fortunate to bring the Pohang plant's First Stage on stream during the world-steel boom in 1973. However, by the time the Second Stage was completed in 1976, the boom had gone bust, leaving steel in surplus supply around the world to the detriment of steel prices and steel-company earnings. Far from being deterred by this hostile turn of events, POSCO immediately commenced Pohang's Third Stage to more than double the plant's steelmaking capacity to 5.5 million annual tonnes. In so doing, the company adhered to another important tenet of its building strategy, that is, to persevere in expanding capacity and growing the business during periods of downturn for the steel industry at large. This strategy has afforded POSCO considerable leverage in negotiating facility costs with equipment suppliers who are begging for work, which has been yet another reason for the company's success in controlling capital costs. The world steel industry's retrenchment from the mid-1970s through most of the 1980's gave POSCO considerable opportunity to employ this aspect of its building strategy.

The head of another successful corporation, the world's largest conglomerate, General Electric Company, concurs in this approach to pursuing capital investments. As stated by GE chairman Jack Welch, "The time to put the pedal to the metal is when everyone else is hunkering down." Doing this, in his view, is the way to get a major leg up on the competition.[2] And this is precisely what POSCO has done.

Pohang's Third Stage, representing an investment of $469 per tonne, was completed within 29 months and dedicated on December 8, 1978. Shortly thereafter, construction was started on the Fourth Stage, which was designed to raise capacity to 8.5 million annual tonnes. This was accomplished within 25 months, and the Fourth Stage was dedicated on February 18, 1981, whereupon a second phase of the Fourth Stage was initiated to again increase capacity, this time to 9.1 million tonnes. This "final" addition to the Pohang plant was dedicated on May 25, 1983, and since that time, the plant's capacity has trended incrementally higher to a current level of 12.7 million tonnes, the added capacity coming both from productivity improvements and the construction of new facilities, including two electric steelmaking furnaces for stainless-steel production.

Each of the stages of the Pohang plant's construction complied with the rule of being completed on or ahead of schedule, which, along with the other tenets of POSCO's building strategy, succeeded in holding overall capital costs at the conclusion of the Fourth Stage to $3.8 billion, representing an extremely favorable average of $422 per tonne, much lower than the estimated worldwide average at the time of $600 to $800 per tonne. This helped POSCO become highly competitive in international markets, which together with the company's intensive employee training and education programs and its securing of stable sources of raw materials, went a long way in laying the foundation for continuous profitability and the internal funding of future expansion projects at considerable cost savings[3].

The Pohang plant is now in its 28th year of operation and has been continuously modernized and upgraded to remain one of the world steel industry's largest, most highly regarded, and competitive integrated steelworks. Despite its size, Pohang is distinguished by its ability to efficiently produce relatively small volumes of a wide variety of high-quality finished steel products, presently including hot- and cold-rolled sheets, plates, wire rods, galvanized sheets, tinplate, stainless steel sheets, and silicon-steel sheets.

The plant encompasses some 40 production facilities and a near equivalent number of support facilities, the most significant of the producing units being four blast furnaces (See Figure 2.1), one foundry blast furnace, one COREX ironmaking unit, five coke-oven batteries, five sinter plants, two oxygen steelmaking shops, three continuous-casting plants, two hot-strip mills, two cold-rolling mills, one electrolytic tinning line, one electrolytic galvanizing line, three plate mills, three wire-rod mills, one silicon-steel facility, and two stainless-steel facilities with two electric-arc furnaces. The capacities and completion dates of the major production facilities at Pohang are shown in Table 2.2.

As presently configured, the Pohang plant is divided into two roughly equivalent sections, separated by a straight road known as Central Avenue. The plant's hot end, including its blast furnaces and steelmaking facilities, is situated on one side of the road, and its rolling and finishing facilities are on the other.

Central Avenue runs toward the harbor area, which has two raw-materials unloading docks. The largest is located at the outer harbor and can simultaneously berth one 250-thousand-tonne bulk carrier, another of 150 thousand tonnes, and three of 100 thousand tonnes. The unloading dock at the inner harbor can accommodate two 50-thousand-tonne carriers. The two docks have a combined capacity to unload 15 thousand tonnes per hour for deposit at the plant's raw-materials storage yard prior to being transported by conveyor for subsequent use.

Supporting the Pohang plant's blast-furnace operations are five coke batteries with a combined annual capacity of 5.0 millions tonnes, a coal-chemicals facility to process coke-oven gas, and five iron-ore sintering plants with a total processing capacity of 14.7 million annual tonnes. In addition, a COREX ironmaking unit supplements Pohang's blast-furnace capabilities by smelting up to 700 thousand annual tonnes using coal rather than the coke needed to make blast-furnace iron.

Pohang has four blast furnaces that can smelt a combined 9.2 million annual tonnes of iron for steelmaking and one foundry blast furnace with a yearly capacity to produce 810 thousand tonnes of pig iron for the manufacture of castings (see Table 2.3). Dating back to the 1973 startup of

Figure 2.1 The Pohang Plant's Blast Furnaces

Table 2.2 Facility Capacity of Pohang Works

Facility	*Plant*	*Completion Date*	*Capacity (thousand MT)*
Sintering	#1 Sinter	1973.5	1,857
	#2 Sinter	1976.5	2,663
	#3 Sinter	1978.10	4,580
	#4 Sinter	1981.1	5,081
	Foundry Suter	1974.10	543
	Total	—	14,724
Coke	#1 Coke	1973.4	516
	#2 Coke	1976.4	804
	#3 Coke	1978.11	1,436
	#4 Coke	1981.2	1,476
	#5 Coke	1983.6	783
	Total	—	5,015
Blast Furnace	#1 BF	1973.6	1,180
	#2 BF	1976.5	1,900
	#3 BF	1978.12	3,070
	#4 BF	1981.2	3,070
	Foundry BF	1974.10	810
	COREX	1995.11	700
	Total	—	10,730
Steelmaking	#1 Steel	1973.6	2,717
	#2 Steel	1978.12	8,718
	Total	—	11,435
Continuous Casting	#1 CC	1976.5	2,600
	#2 CC	1981.2	4,200
	#3 CC	1986.6	4,200
	Total	—	11,000
Hot Rolling	#1 HR	1972.10	2,709
	#2 HR	1980.7	4,219
	Total	—	6,928
Plate	#1 Plate	1972.7	539
	#2 Plate	1978.2	1,808
	#3 Plate	1997.9	1,157
	Total	—	3,504
Billet	Billet Made of Bloom	1972.6	772
Wire Rod	#1 WR	1979.2	705
	#2 WR	1984.11	467
	#3 WR	1988.11	780
	Total	—	1,952
Cold Rolling	#1 CR	1977.2	1,151
	#2 CR	1987.2	1,200
	Total	—	2,351
Silicon Steel	Electrical Steel	1979.10	508
Stainless Steel	STS Steelmaking	1989.3	1,240
	CC	1989.3	1,200
	HR	1989.3	1,076
	CR	1990.7	171

Source: Pohang Iron and Steel Company, Ltd.

Table 2.3 The Pohang Plant's Blast Furnaces

	No. 1	No. 2	No. 3	No. 4	Foundry
Startup Date	1973	1976	1978	1981	1990
Hearth Diameter (m)	9.2	11.0	13.2	13.2	7.7
Inner Volume (m^3)	1660	2550	3795	3795	1080
Iron Capacity (t/d)	3233	5205	8411	8411	2219

Note: m = meter, t/d = tonnes per day.
Source: Pohang Iron and Steel Company, Ltd.

ironmaking operations at Pohang, all of the plant's blast furnaces have been continuously modernized to receive the latest in process technologies and practices, including pulverized coal injection (PCI) and computerized automatic control systems. Frequent modernizations have periodically extended the size and capacity of the furnaces and lowered their unit costs of production.

The Pohang plant's two oxygen steelmaking shops house a combined six oxygen converters. No. 1 shop has three 100-tonne vessels rated at 2.7 million annual tonnes, whereas No. 2 shop has three 300-tonne vessels capable of producing 8.7 million tonnes. The two steelmaking shops have degassing and refining equipment to treat most of the steel they produce.

The liquid steel is processed into semifinished form at three continuous-casting shops housing a total of 10 continuous-casting machines. Two billet casters with a combined 14 strands can process up to 1.3 million tonnes annually in sizes from 120 to 200mm square. A single bloom caster with four strands can process 650 thousand tonnes of blooms 250 × 330 mm. And five slab casters with a total of nine strands have a combined capacity to cast 9.0 million tonnes of slabs with a thickness range of 200 to 250mm in widths ranging from 740 to 2200mm.

On the opposite side of Central Avenue from where Pohang's above-described hot end is situated, an orderly array of mill buildings houses the plant's numerous rolling and finishing facilities. Pohang's hot-strip-mill complex contains two continuous, wide hot-strip mills. The first is 1270mm wide, has one roughing stand and seven finishing stands, and can process some 2.7 million annual tonnes of hot-rolled coils. The second is 1880mm wide, has four roughing stands in advance of its seven finishing stands, and can roll 4.2 million annual tonnes.

Some 2.4 million tonnes of Pohang's hot-rolled coils can be cold rolled on two cold-reduction mills. The first is 1270mm wide, has four four-high stands and one six-high, and is capable of rolling in excess of 1.1 million annual tonnes. The second is 1650mm wide, has four six-high stands, and is rated at 1.2 million annual tonnes. Completed in 1987, the No. 2 cold-rolling mill, which is linked to a continuous annealing line and an electrolytic-galvanizing line, delivers sheet products to meet the growing demands of Korea's automobile industry. One cold-rolling mill for silicon steel produces 508 thousand tonnes of electrical sheets a year.

The plant's three rod mills have a combined capacity to roll some 2.0 million annual tonnes of high-quality rods in a wide range of sizes and carbon or alloy grades to meet the needs of customers manufacturing fasteners, springs, wire rope, welding rods, and other wire products. Pohang's three plate mills are capable of rolling a combined 3.6 million annual tonnes of high-grade plate products. Completing the roundup of Pohang's rolling facilities, all the plate mills are four-high reversing type and have been instrumental in facilitating the growth of Korea's shipbuilding industry. Since their startup in 1972 and 1978, No. 1 and No. 2 plate mills have been upgraded regularly to ensure the highest plate quality, while construction of the No. 3 plate mill was completed in September 1997.

Finally, the stainless steel plant houses two electric-arc furnaces for melting stainless steel, one a 90-tonne, 70MVA unit and the other a 90-tonne, 85MVA unit, which have a combined capacity of 1.23 million annual tonnes. Two casters can process 1.2 million tonnes of stainless-steel slabs annually. The STS APL (Annealing and Pickling Line) can treat 1.07 million tonnes of black coils. STS cold mills at Pohang are designed to process some 171 thousand annual tonnes of stainless sheets.

All of the Pohang plant's facilities have been regularly modernized to incorporate the latest in technological advances as they become available. A key aspect of the POSCO strategy has been to recognize that the building process never ends, because the requirement for plant and equipment modernization must be pursued on a continuous basis to stave off the onset of creeping obsolescence to which many other steelmakers have fallen prey. POSCO's ability to sustain profitable operations, an objective that has eluded others, has been critically important to sustaining the modernization process.

Today, the Pohang plant ranks as one of the largest integrated steelworks in the world in terms of its steelmaking capacity, and POSCO is the only steel company in the world with two integrated plants in the top 10. POSCO's second integrated plant at Kwangyang, now the world's largest, was envisioned by the company as "The Steel Mill of the 21st Century," and with its construction, POSCO has become the largest steel-producing company in the world.

THE KWANGYANG PLANT

During 1978, when POSCO was working toward completion of the Pohang plant's Third Stage, considerable discussion arose within Korea's government and business circles about constructing another integrated steelworks to give added steel support to the nation's rapidly growing economy. On October 27 of that year, Korean President Chung-Hee Park, mindful of the remarkable success achieved at Pohang, selected POSCO to build the second plant, and some three years later, on November 4, 1981, his successor, President Doo-Hwan Chun settled a protracted controversy over where the plant should be located by selecting a site on Kwangyang Bay that POSCO had strongly supported.[4]

By the end of 1981, the case for constructing the second integrated plant had become increasingly clear. Korea's automobile, shipbuilding, and other manufacturing industries had entered their take-off stages and were experiencing strong production growth, with corresponding increases in their steel requirements. In response, POSCO had amended the Pohang plant's Fourth Stage, and with government planners calling for steel demand to continue on a strong growth track in the decade to come, POSCO also had set in motion the planning process for building a plant at Kwangyang to more than double the company's steelmaking capacity.

Most of the Pohang plant's expansion and the ramping up of POSCO's steel output had coincided with Korea's third and fourth five-year economic plans for the years 1972–76 and 1977–81. During these two periods, due in significant part to the stimulus and support provided by POSCO's unprecedented growth and development, Korea's yearly economic growth rates had averaged 10.1 percent and 5.6 percent, even though growth in the latter period had been impacted by the 1979 world oil shock. Over the same five-year periods, steel demand, reflecting the rapid onset of industrialization, had grown at yearly averages of 16.7 percent and 13.5 percent, representing an extremely strong performance.

To further advance Korea's industrialization during the period through 1991, spanning the fifth and sixth five-year economic plans, the nation's Economic Planning Board had established growth targets for the economy and for steel consumption averaging 7.3 percent and 8.4 percent per year, respectively. This was seen as leaving Korea with a mounting deficit in the supply of domestically produced steel, unless a second integrated plant, at least on the scale of that at Pohang, was to be constructed.[5]

A primary consideration in President Chun's selection of the site at Kwangyang was the opportunity the location afforded for balanced regional economic development by attracting industrial activity into the Honam area, west of the city of Pusan in Chun Nam Province at the southern end of the Korean peninsula. From POSCO's standpoint, Kwangyang was attractive because of its favorable climate, minimal tidal shifting, seas that most often were placid, particularly when compared to those at Pohang, and a deep-water shoreline that was conducive to constructing port facilities for berthing the largest bulk-cargo carriers.

However, the Kwangyang site also presented some difficult challenges from an extensive weak soil formation that was prone to greater settlement than could be tolerated for steel-plant construction. Primary responsibility for overcoming this problem was assigned to Sang-Boo Yoo, who today is POSCO's chairman. The team he headed placed the cost of stabilizing the site at $150 million, substantially less than the $1.0 billion estimate made by foreign consultants for Korea's Construction Ministry. But at the conclusion of 33 months of site preparation, the job was completed at the exceptionally low cost of $100 million.[6]

Prior to starting site preparation work at Kwangyang, POSCO conducted extensive land surveys, soil testing, and laboratory simulations. A priority of the overall analysis was to protect the surrounding natural environment, including that at nearby Hallyo Waterway National Park. The findings revealed that the Kwangyang site's total area of 3,700 acres was characterized by weak soil formations to varying depths and that the sand-covered surface was subject to settling in a range of 70 to 270 centimeters (roughly two through nine feet). The findings also established the most economical and technically suitable method for strengthening the weak soil, namely hydraulic fill and compaction, then a new technological approach that had been used with success by steelmakers in Japan.

On September 28, 1982, POSCO started to reclaim the land at Kwangyang. The first step was constructing a 13.6 kilometer-long system of dikes, a bracket-shaped wall in the sea to the immediate south of Kumho Island. The project employed 280 thousand workers and 78 thousand pieces of heavy equipment and, when completed, defined the perimeter of what would become a man-made island on which the company's second steelworks was to be built.

As dike construction progressed, dredging and reclamation work simultaneously was carried out, with silt protection used to safeguard adjacent waterways. Next, the process of weak soil improvement was implemented to stabilize and strengthen the site down to bedrock at depths of 20 to 25 meters. For the first time in Korea, a system of vertical drains, spaced at intervals of 2 to 2.5 meters, was used to drive excess water and air from the underground soil prior to backfilling and compaction. In total, an area of 14.9 square kilometers had to be filled to grade, and some 220 thousand mounds of sand and soil had to be driven into the ground, their combined volume exceeding 11 million cubic meters. This massive site reclamation, representing a major engineering achievement, progressed ahead of schedule, permitting the First Stage of the Kwangyang plant's construction to commence on March 5, 1985. The site work itself was completed on June 30, 1985.

In building the Kwangyang plant, POSCO drew considerable benefit from the technological know-how and experience gathered over the prior years of constructing and operating the plant at Pohang. Building plans for both plants provided for a four-stage construction, which in the Kwangyang plant's case was designed to add 2.7 million annual tonnes of steelmaking capacity in each of the first three stages and 3.3 million tonnes more in the fourth and final stage. As originally conceived, this was to provide POSCO with an aggregate capacity addition of 11.4 million annual tonnes, which has since been amended incrementally to give Kwangyang its present steelmaking capacity of 16.5 million tonnes.

The First Stage of Kwangyang's construction was completed in 26 months and dedicated on May 7, 1987. It encompassed 10 major production installations and 14 auxiliary facilities, the former including a coke plant, sinter plant, blast furnace, basic-oxygen steelmaking shop, continuous-casting plant, and hot-strip mill, and the latter including all of the plant's necessary utilities and service equipment, port and berthing facilities, the plant office building, and an employee housing development on Kumho Island to provide a quality living environment for 2,208 households. POSCO's investment in the First Stage, accounting for the expenditures needed to build Kwangyang's infrastructure, totaled $1.95 billion, representing an average capital cost of $723 per tonne of steelmaking capacity. Of the total cost, 70.7 percent was financed domestically by POSCO, with the balance obtained through overseas loans.

Kwangyang's First Stage underscored a decision made by POSCO to implement iron and steel technologies from a much more diverse group of international suppliers than had been employed in the initial stages of the Pohang plant's construction, which of necessity had relied much more heavily on equipment purchases from Japan. In approaching Kwangyang's construction a decade or so later, POSCO strategically sought to ensure that the plant would be equipped with the most advanced facilities and technologies available anywhere by purchasing equipment from suppliers in a number of major steelmaking countries, including Japan, the United Kingdom, Germany, Austria, and France. In addition, although Korean suppliers had been relied upon for no more than 12.5 percent of the facilities that went into the Pohang plant's First Stage, the proportion of local facilities used in constructing Kwangyang's First Stage reached 49.4 percent, representing clear evidence of the extent to which POSCO's growth and development had promoted Korea's industrialization.[7]

During the decade preceding the start of Kwangyang's construction, difficult times for most of the world steel industry had made retrenchment the order of the day. However, the state of the art in iron and steel technology had continued to advance, with innovations made as much by builders of steel-mill equipment as by steel-producing companies. The innovations not only took the form of process improvements from ironmaking through rolling and finishing, but also involved the application of evolving computer technologies to more effectively control and coordinate the myriad of processes and activities involved in steel-plant operations. POSCO's strategy had been to adopt many of these technological advances in continuously modernizing the Pohang plant, and the company pursued their fullest implementation in building Kwangyang with the objective of eventually making "The Steel Mill of the 21st Century" a commercial reality.

Toward this end, the Kwangyang plant was designed to continuously cast 100 percent of its raw-steel output, to employ hot direct rolling of its semifinished output, and to more closely control the shape of its finished steel products. Further, an on-line, real-time computer information system was installed to integrate all of the plant's production processes and all aspects of its business, from operations to management and finance, into a single network. The computer system was programmed in large part by POSCO employees to afford on-line control of all manufacturing activities, from the receipt of raw materials to the shipment of finished products, and proved to be a key contributor to increasing productivity and reducing production costs. Although comparable computer systems and process technologies have become increasingly common in today's leading steel plants, POSCO's successful implementation of these advances at an early date gave the company a decisive technological lead over its competitors in other countries.

In building the Kwangyang plant, POSCO consistently met its standard for on-time execution by completing each step of the four-stage construction two to six months ahead of schedule. This had a positive impact on investment costs and also served to make timely capacity additions in attempting to meet Korea's growing demands for steel and thereby alleviate a mounting domestic supply deficit. Given the urgency to boost supply, work on Kwangyang's Second Stage expansion commenced on September 30, 1986, when construction of the plant's First Stage was still some seven months from completion. The Second Stage, designed to double the plant's steelmaking capacity to 5.4 million annual tonnes, required an investment of $1.28 billion, or $473 per tonne, and was completed within 22 months and dedicated on July 12, 1988.

To further address Korea's growing steel needs, Kwangyang's Third Stage expansion was started on November 1, 1988, and, upon completion 25 months later on December 4, 1990, the plant's capacity was increased to 8.1 million annual tonnes. The Third Stage placed major emphasis on the installation of finishing facilities, including additions to an advanced cold-rolling-mill complex, which was reflected in an investment requirement of $2.83 billion, or $1,048 per tonne of added steelmaking capacity.

Shortly after the Third Stage was completed, work on Kwangyang's final Fourth Stage commenced on January 5, 1991, aimed at installing an additional 3.3 million annual tonnes of capacity. The Fourth Stage was completed within 21 months, on October 2, 1992, raising Kwangyang's total steelmaking capacity to 11.4 million annual tonnes and concluding the largest single expansion project in POSCO's history. Investment requirements for the Fourth Stage were $2.51 billion, or $761 per tonne, resulting in an overall capital cost for Kwangyang's four-stage construction of $8.57 billion, or an average outlay of $752 per tonne.

Over the nine or so years since the Fourth Stage was completed, POSCO has adhered to the strategy of continuous modernization in upgrading Kwangyang, and the plant's steelmaking capacity has increased incrementally to its current level of 16.5 million annual tonnes. Part of the increase in capacity has come from process and practice improvements and productivity gains at facilities already in place, and part from the installation of such additional facilities as another blast furnace, and even a thin-slab minimill to bolster sheet production. The major production facilities now in place at Kwangyang are listed in Table 2.4, along with their completion dates and annual capacities.

As presently configured, the Kwangyang plant employs a straight-line layout conducive to the direct, computer-controlled linkage and coordination of its primary plant and equipment components. This feature, as much as any other, gives Kwangyang its "21st century" qualities and contributes significantly to the plant's high productivity, energy efficiency, and competitive production costs. By directly connecting the plant's various operating functions, POSCO has succeeded in reducing the time required to convert molten iron into shippable steel products to less than four hours, instead of the four or more days that were once the standard in most integrated steel plants.

Kwangyang's operations start at the largest port facility in Korea, featuring a four-kilometer-long unloading wharf, a water depth of 22 meters, and the capabililty to simultaneously berth five large vessels, including a maximum bulk cargo carrier of 250 thousand deadweight tonnes. Iron ore, coking coal, and other raw materials are unloaded from the ships at 14 thousand tonnes per hour, are deposited in an expansive stockpile area where stringent dust-abatement and other environmental-control procedures are adhered to, and are subsequently transferred by conveyor for processing at Kwangyang's sinter plants, coke ovens, and blast furnaces. West of the unloading wharf are load-out berths for shipping finished steel products in vessels of some 5–50 thousand deadweight tonnes, and lo-

Table 2.4 Facility Capacity of Kwangyang Works

Facility	*Plant*	*Completion Date*	*Capacity (thousand MT)*
Sintering	#1 Sinter	1987.4	4,765
	#2 Sinter	1988.7	4,765
	#3 Sinter	1990.11	4,765
	#4 Sinter	1992.9	5,051
	Total	—	19,346
Coke	#1 Coke	1987.4	1,488
	#2 Coke	1988.7	1,508
	#3 Coke	1990.12	1,508
	#4 Coke	1990.9	1,508
	Total	—	6,012
Blast Furnace	#1 BF	1987.4	3,066
	#2 BF	1988.7	3,006
	#3 BF	1990.12	3,066
	#4 BF	1992.9	3,066
	#5 BF	1999.3	3,066
	Total	#5 BF operation started in 2000.4	15,330
Steelmaking	#1 Steel	1987.4	7,178
	#2 Steel	1990.11	7,565
	Total	—	14,743
Continuous Casting	#1 CC	1987.3	6,800
	#2 CC	1990.11	7,200
	Total	—	14,000
Hot Rolling	#1 HR	1987.2	4,892
	#2 HR	1990.8	3,634
	#3 HR	1992.7	3,984
	Minimill	1996.10	1,800
	Total	—	14,310
Cold Rolling	#1 CR	1989.1	1,679
	#2 CR	1991.1	1,663
	#3 CR	1992.9	1,005
	#4 CR	1997.8	2,195
	Total	—	6,542

Source: Pohang Iron and Steel Company, Ltd.

cated to the east is the port's Central Terminal System, where coal received in 150-thousand-tonne ships is stored and later delivered using an efficient system of coastal barges to major coal consumers throughout Korea.

Moving inland from Kwangyang's port, cokemaking and sintering operations are conducted at locations immediately behind the raw-materials stockpile area and in front of the plant's ironmaking section with its array of large blast furnaces (See Figure 2.2). At the cokemaking facility, blends of more than 40 kinds of coal are carbonized in four 132-oven coke batteries, which have a combined annual capacity of 6.0 million tonnes, and four iron-ore sintering lines are capable of processing a combined 19.3 million annual tonnes of sinter.

The Kwangyang plant's five blast furnaces are comparable in size to the largest blast furnaces at the Pohang plant, all having hearth diameters of 13.2 meters and inner volumes of 3,800 cubic meters (See Table 2.5). Together, the five furnaces have the capacity to produce 15.3 million annual tonnes of molten iron for subsequent processing into steel. In order to maximize the quantity and quality of the iron produced and minimize iron-ore and coke consumption, the furnaces are equipped to employ the most advanced smelting techniques, including fully prepared burdens, optimum wind rates and blast temperatures, bell-less charging, pulverized coal injection (PCI), and hot-metal pretreatment for silicon, sulfur, and phosphorous removal. Top gas from the furnaces is recovered for use in a turbine-generator power station to provide a source of electricity for use throughout the plant.

The No. 1 through No. 4 blast furnaces were installed during Kwangyang's four-stage construction, whereas No. 5 furnace went into operation in April 2000, its installation having been related in part to the need for supplemental iron as a feedstock for the electric steelmaking furnaces at Kwangyang's thin-slab minimill, which had been completed in October 1996. The electric furnaces refine the iron in conjunction with the melting of high-quality revert scrap generated within the Kwangyang plant, which, given the plant's high production yields, is short of the minimill's total charge requirements.

The steelmaking section at Kwangyang houses two basic-oxygen (BOF) converter shops for producing a variety of high-quality molten steels. Each shop has three 250-tonne per heat steelmaking vessels and is equipped with combined blowing technology, computerized dynamic process control, and secondary ladle refining and degassing capabilities. The two BOF shops have a combined steelmaking capacity of 14.7 million annual tonnes, and the Kwangyang minimill has the capacity to melt an additional 1.8 million annual tonnes. The minimill's steelmaking shop

Figure 2.2 The Kwangyang Plant's Blast Furnaces

Table 2.5 The Kwangyang Plant's Blast Furnaces

	No. 1	*No. 2*	*No. 3*	*No. 4*	*No. 5*
Startup Date	1987	1988	1990	1992	2000
Hearth Diameter (m)	13.2	13.2	13.2	13.2	13.2
Inner Volume (m^3)	3800	3800	3800	3800	3800
Capacity (t/d)	8400	8400	8400	8400	8400

Note: m = meter, t/d = tonnes per day.
Source: Pohang Iron and Steel Company, Ltd.

operates two 130-tonne per heat, twin-shell, DC electric-arc furnaces and two ladle-refining furnaces.

Kwangyang has two continuous-casting complexes to process liquid steel from its BOF shops exclusively into slab form. The two complexes house a combined eight slab-casting machines with a total of 16 strands and a capacity to process 14.0 million annual tonnes of slabs having a thickness of 230mm in widths ranging from 820 to 2000mm. Electric-furnace steel produced at Kwangyang's minimill is processed on two, single-strand machines with reduction units into so-called thin slabs with a thickness of 25 to 30mm in widths ranging from 900 to 1350mm.

The hot-strip-mill complex at Kwangyang contains three continuous, wide hot-strip mills with a combined capacity to process 12.5 million annual tonnes of hot-rolled coils. No. 1 mill is 1650mm wide, incorporates three roughing stands and eight finishing stands, and has an annual capacity of 4.8 million tonnes. No. 2 mill, which is 1570mm wide and has two roughing and six finishing stands, can process 3.6 million annual tonnes, whereas No. 3 mill is 1950mm wide, has two roughing and seven finishing stands, and can roll close to 4.0 million annual tonnes. All three mills are equipped with the latest in rolling technologies, including full computer control, direct hot-charge rolling or HCR, and pair-cross rolls for enhanced shape and crown control, which are essential to attaining high product quality. Kwangyang's minimill has a hot-strip mill as well. It has five finishing stands, is 1480mm wide, and can roll 1.8 million annual tonnes.

Four cold-reduction mills at Kwangyang have a combined capacity to roll 6.5 million annual tonnes of cold-rolled coils. All four mills are the six-high type and supply product to downstream annealing, galvanizing, and other processing facilities. No. 1 and No. 2 cold mills both have four stands, and each has a rolling capacity of just under 1.7 million annual tonnes. The former has a maximum width of 1880mm and the latter 1570mm. No. 3 cold mill has five stands, a maximum width of 1270mm, and is capable of rolling 1.0 million annual tonnes.

No. 4 cold mill, which was completed in August 1997, is among the world's most advanced six-high cold-reduction mills, having been designed to meet the most exacting demands of the automobile and appliance industries. A five-stand tandem mill, it has an annual rolling capacity of 2.2 million tonnes and can process sheets in the range of 0.4 to 2.3mm thickness, 720 to 1890mm in width, and in high tensile strengths of the 35kg and 45kg class needed for auto-body manufacturing. The mill is linked to two million-ton-plus continuous annealing lines and an electro-galvanizing line, the former having the capability to treat ultra-wide sheets at a rate of 430mm per minute. No. 4 mill incorporates an automatic gauge control system, an artificial intelligence system, and other state-of-the-art rolling technologies developed by POSCO, which combine to deliver a finished product with exacting dimensions of thickness, width, and length. The mill's product capabilities complement those of the five other cold-reduction mills within the company's two integrated plants.

THE BUILDING CONTINUES

POSCO's accomplishment since starting to transform an underdeveloped tract of land at Pohang represents an extraordinary chapter in the history of modern industry. The success achieved in building two of the world's leading integrated steel plants, including the world's largest and most advanced at Kwangyang, demonstrates that the process of building a viable steel company never ends. The same strategies that were successfully employed to execute the original four phases of each plant's construction have remained in place to advance the ongoing process of modernization and expansion.

Since the original capacity objectives for the Pohang and Kwangyang plants were attained, POSCO has increased their combined steelmaking capabilities by 8.7 million annual tonnes. Over the past 30 years, dating back to the start of Pohang plant's construction, POSCO's consistent pursuit of a proven building strategy has given the company a current steel-making capacity, including stainless, of 29.2 million annual tonnes. In the process of building this capacity, POSCO spurred the building of a nation and today stands as a powerful catalyst for Korea's continued economic and industrial development.

NOTES

1. J.J. Innace and A.D. Dress, *Igniting Steel* (New York: Global Village Press, 1992), pp. 44–45, and K.K. Seo, *The Steel King* (New York: Simon & Shuster, 1997), pp. 123–124.
2. J. Welch, Interview, *Charlie Rose,* PBS Television, March 16, 2001.
3. H.E. Kim, ed., *A Miracle on Yongil Bay* (Seoul: Pohang Iron and Steel Company, Ltd., (1983), pp. 9–10.
4. Pohang Iron and Steel Company Ltd., *Dedication of Kwangyang Steelworks* (Seoul: Pohang Iron and Steel Company, Ltd., May 1987), p. 7; and Seo, *op. cit.*, pp. 338 and 345.
5. Pohang Iron and Steel Company, Ltd., *The Necessity of the Construction of Kwangyang Works from a National Economic Point of View*, Company Analysis, March 1983, pp. 2–4 and 31.
6. Seo, *op, cit.*, pp. 344–348.
7. Kim, *op cit.*, p. 10.

3

The Market Strategy

Since the mid-1960's, just prior to the time POSCO was founded, the Korean steel market has undergone a more than 100-fold growth, from just 328 thousand tonnes in 1965 to 38.5 million tonnes in the year 2000. Once steel from POSCO's first integrated steel plant started to work its way through Korea's economy, a truly remarkable period of economic and industrial development started to unfold. The growth of Korea's steel-consuming industries over the ensuing quarter century would not have been possible without the stimulus and essential support provided by the growth of domestic steel production, most of which was derived from POSCO's success in building and continually upgrading two of the world's largest and most competitive integrated steel plants.

The Korean steel market's growth during 1975–2000 was powered by and closely paralleled the growth of domestic crude-steel production and the rise in POSCO'S tonnage sales of finished steel products (See Table 3.1). Korea's steel market, gauged in terms of apparent consumption, namely domestic shipments plus imports minus exports, expanded at a compound annual rate of 11.8 percent, from 2.4 million tonnes in 1975, when POSCO's Pohang plant entered its third year in operation, to the aforementioned 38.5 million tonnes in 2000. In order to satisfy this growth in demand, the Korean steel industry increased its output of crude steel at a 12.0 percent yearly rate, from 2.6 million to 43.1 million annual tonnes, and POSCO boosted its finished-product sales an average of 13.5 percent each year, from 1.1 million to 26.8 million tonnes.

Just as Korea's industrialization and the growth of its steel-consuming

Table 3.1 Korean Apparent Steel Consumption, Crude Steel Production, and POSCO Sales of Finished Steel Products: 1975–2000
(thousands of tonnes)

	Apparent Consumption	*Crude Steel Production*	*POSCO Sales of Finished Steel*
1975	2,371	2,558	1,118
1976	3,174	3,511	1,703
1977	4,719	4,346	2,275
1978	6,042	4,969	2,742
1979	6,204	7,610	4,239
1980	5,115	8,558	5,018
1981	6,781	10,753	6,757
1982	6,492	11,758	7,682
1983	7,801	11,915	7,707
1984	9,408	13,034	8,068
1985	10,020	13,539	8,320
1986	10,934	14,555	8,841
1987	13,642	16,782	10,928
1988	14,519	19,118	12,814
1989	16,946	21,873	14,797
1990	20,054	23,125	15,739
1991	24,454	26,001	18,234
1992	21,820	28,055	19,237
1993	25,246	33,026	21,183
1994	30,510	33,745	21,839
1995	35,529	36,772	22,899
1996	37,583	36,903	23,601
1997	38,146	42,554	25,171
1998	24,720	39,896	24,915
1999	33,761	41,042	25,737
2000	38,467	43,107	26,783

Source: Pohang Iron and Steel Company, Ltd., Korean Iron and Steel Association, and International Iron and Steel Institute.

industries could not have taken place without a thriving domestic steel industry, POSCO's transformation from a developing-country startup into the world's foremost steel-producing company could not have taken place without the exceptional growth of the Korean steel market. The early recognition of this industrial interdependence by Korea's policymakers had made construction of an integrated steel plant the focus of government efforts to foster economic development, and POSCO, in turn, effectively advanced this objective by pursuing a consumer-driven market strategy.

Considering that POSCO has spent much of its history striving to produce enough steel to keep pace with the Korean market's burgeoning needs, the question arises as to just how important the company's market strategy has been. If the company has had waiting buyers for all the steel it can make, then why bother to have a market strategy in the first place? The answer is that if Korea's once fledgling industries were to prosper and become major steel consumers, they first had to become world-class manufacturers of such products as supertankers, automobiles, and household appliances, and to do so, they required not just steel, but certain types of high-quality steel. A consumer-driven market strategy has made it possible for POSCO to effectively determine and satisfy the specific and growing needs of Korea's many steel-consuming industries.

When POSCO started to produce steel in mid-1973, the Korean steel industry, limited by poor facilities and backward technologies, could supply only one-third of the domestic steel market's needs. Korean steel consumers were disadvantaged by having to rely on overseas sources to meet their requirements for high-quality steel. By the mid-1970's, POSCO's ability to supply this steel at low cost was undergoing the first of many expansions that were to boost the company's steelmaking capacity nearly 30-fold. In the process, the company became the engine that accelerated the development of Korea's steel-consuming industrial base, which, in turn, uplifted the overall national economy.

Well before POSCO sold its first tonne of steel, the company had closely evaluated the needs of Korea's steel consumers to guide the selection of facilities for the company's first integrated plant. Recognizing the market's need for a wide variety of steels, the Pohang plant was given diverse capabilities to produce both long and flat products and was designed to achieve a high degree of operating efficiency even while processing many individual orders in relatively small volumes. By the time construction of POSCO's second plant at Kwangyang commenced in 1985, Korea's industrial base was well into its takeoff stage, creating a surge in the domestic market for high-quality steel sheets. Once again, POSCO responded to the market's need by designing Kwangyang to engage exclusively in the high-volume production of sheets. According to the company, its intention in building Kwangyang was to "set up a system of one company with two works, which enables us to pursue less quantity but greater product variety at the Pohang Works and vice-versa in the Kwangyang Works."[1]

Operating two decidedly different integrated plants has afforded POSCO the flexibility to meet varying trends in market demand to better accommodate the growth and development of Korea's steel-consuming industries. Accommodating the company's customers has long been an integral part of day-to-day plant operations, as opposed to being reserved primarily for sales and marketing contact. In this manner, the company's market strategy relies on active employee involvement at all operating and managerial levels to ensure that customer satisfaction is firmly embedded in the corporate culture.

Widespread employee involvement in POSCO's marketing effort is a natural outgrowth of a longstanding commitment to employee education and the realization that innovative ideas, including those on how to better serve the customer, are to be found throughout the company. With total customer satisfaction established as an objective critical to the company's long-term viability, employee suggestions on enhancing customer service are actively sought, considered, and regularly acted upon.

According to POSCO, customer-satisfaction management is "one of the most important management objectives of marketing," and in conducting employee education programs in serving the customer, it is the company's firm belief that "customer satisfaction begins with the minds of its employees."[2]

POSCO's market strategy makes total customer satisfaction a goal synonymous with the company's striving to achieve and maintain the distinction of being the world's top-quality steel producer. To concurrently serve both objectives, true partnerships are formed with customers by means of the company's Mill Representative System, which is designed to customize processing technologies within the Pohang and Kwangyang plants to meet the specific product requirements of customers, who are also provided with on-site technical support and consultations. Steel orders requiring the use of high technology or calling for exceptional specifications are welcomed, with the objective of increasing their share of the total product mix. By these and other methods, POSCO's market strategy seeks to jointly ensure customer satisfaction and superior product quality and thereby contribute to securing a competitive advantage in the marketplace.

Total customer satisfaction has also required making the steel-ordering process more customer oriented, mainly by continually working to reduce the average time required to complete an order, from placement to final product shipment. In facilitating sales, credit and payment terms have been made more flexible and diversified, and the company's product distribution network has been continually strengthened, such as by developing "Roll on–Roll off" shipping docks at Kwangyang and Pohang to improve the coastal marine transport of the company's steel.

Increasingly, POSCO's market strategy is being focused on maximizing the production and sale of high-value-added steel products, not only by actively responding to changes in market demand, but by working to shape those changes through an extensive product-development effort. In seeking to create new demand for its products, the company has developed a steel beverage can to promote recycling, has been promoting the use of steel in residential-building construction, and also has actively participated in an international program to design lighter, more fuel-efficient automobiles. In addition, the company has been a supporting member of organizations formed by the Korean Iron and Steel Association (KOSA) to create new sources of steel demand. These organizations include the Steel House Club, the Steel Construction Club, and the Stainless Steel Club.

The many and diverse initiatives that comprise POSCO's market strategy are, in part, a reflection of the need to serve customers in diverse industries, wherein steel is used in a seemingly endless variety of manufactured and other products, ranging from food and beverage containers to high-rise buildings. The impact that POSCO has had on Korea's industrial development can be discerned from a discussion of the major Korean steel-consuming industries that account for most of the demand for the company's products. These industries use steel in automobiles, household appliances, shipbuilding, railroad facilities and equipment, food and beverage containers, oil and gas drilling, pipe and tubing manufacture, and construction. In addition, the steel service centers, which are the retail arm of the steel industry, sell steel in filling relatively small individual orders, although their total sales account for a notably large tonnage.

AUTOMOBILES

The interdependence between POSCO and Korea's automobile industry has long been reflected in the growth and development of both the steel company and one of its foremost steel-consuming markets. In 1968, the year POSCO was founded, Korean motor-vehicle output was less than 18 thousand units. A decade later, as POSCO was completing its fifth year in operation, Korean auto output climbed above the 200-thousand-unit level for the first time, and a decade after that, as the Kwangyang steel plant was at the mid-point of its construction, Korean automakers passed the million-unit milestone and increasingly turned to POSCO to supply the steel needed to drive their continued growth.

The more recent period since 1990 has further reinforced POSCO's ties to Korea's auto-steel market, which is shown by a comparison of the growth in Korean motor-vehicle production and in POSCO's steel shipments to domestic automakers (See Table 3.2). Before the Asian financial crisis impacted Korea's auto and steel industries, the period 1990–97 saw Korean automakers boost their production at a compound annual rate of 10.4 percent. In fact, vehicle output doubled from 1.3 million to 2.6 million units. Notably, POSCO's steel shipments to the automakers through 1997 increased at an even faster rate, averaging 15.8 percent annually, from 1.4 million to a peak level of just under four million tonnes. The greater relative growth of POSCO's auto-steel shipments reflected the company's increasing capabilities to manufacture auto-body sheets to increasingly stringent standards and thereby displace imported steel in

Table 3.2 Korean Motor Vehicle Production and POSCO's Steel Shipments to the Korean Automobile Industry, 1990–1999 (thousands of units and tons)

Year	*Vehicles Produced*	*Steel Shipments*
1990	1,322	1,426
1991	1,498	1,802
1992	1,730	1,480
1993	2,050	2,189
1994	3,312	2,618
1995	2,505	2,165
1996	2,592	3,753
1997	2,641	2,987
1998	1,946	2,417
1999	2,832	3,542

Source: Korea Auto Industries Cooperative Association, *Ward's Automotive Yearbook* (various years), and Pohang Iron and Steel Company, Ltd. Motor-vehicle production includes passenger cars, light and heavy trucks, and buses.

meeting the needs of Korea's auto industry, which now ranks sixth in terms of world motor-vehicle output.

The future growth of Korea's auto-steel market is projected to remain strong over the next 20 years. Forecast figures for motor-vehicle production in 2002 are between three and four million units, and this is projected to increase to seven million units by 2020. Accordingly, the volume of steel consumed by Korea's automakers will increase to 3.9 million tonnes in 2002, to some 4.9 million tonnes in 2010, and to nearly seven million tonnes in 2020, most of which will be in the form of high-quality sheets, the dominant steel product purchased for automotive use.

The significant growth in motor-vehicle production in Korea has been accompanied by a similar growth in motor-vehicle exports, which rose from 347 thousand units in 1990 to 1.5 million units in 1999 (See Table 3.3). Although these exported vehicles are sold throughout the world, their pattern of geographic distribution in recent years has been skewed toward Europe, North America, and Central and South America, where nearly three-fourths of the exports have been marketed, compared to Asia and Oceania, where less than one-fifth have been sold.

In 1999, Korean automakers exported 369 thousand motor vehicles to the United States, 70 percent of which are classified as small cars. By means of highly affordable pricing and long-term warranties, Korean auto manufacturers, namely Hyundai, Kia, and Daewoo have established a dominant position in the entry-level segment of the U.S. car market and

Table 3.3 Korean Motor-Vehicle Exports, 1990–1999 (thousands of units)

Year	*Vehicles Exported*
1990	347
1991	390
1992	456
1993	639
1994	738
1995	978
1996	1,156
1997	1,226
1998	1,362
1999	1,508

Source: Korea Automobile Manufacturers Association and *Ward's Automotive Yearbooks* (various years).

are now starting to upscale their product offerings into the market's mid-to-luxury segments. U.S. sales of Korean light vehicles in 2000 totaled 473 thousand units, representing a sharp 43.6 percent gain from the 330 thousand light-vehicles sold in 1999. Hyundai's 2000 U.S. sales were up 48.8 percent to 244 thousand units, while Kia sold 161 thousand units, representing a 19.3 percent gain, and Daewoo, a relative market newcomer, saw its sales jump 122 percent to 38 thousand units.[3]

Five Korean automakers currently manufacture motor vehicles in significant volumes, namely Hyundai, Daewoo, Kia, Ssangyong, and Samsung. In 1999, they produced a combined 2.6 million passenger cars and trucks, their individual outputs being as follows: Hyundai 1.13 million, Daewoo 736 thousand, Kia 648 thousand, Ssangyong 74 thousand, and Samsung 16 thousand. Table 3.4, which profiles Korea's automobile industry, lists the industry's member companies, together with the locations and capacities of their respective plants that manufacture both passenger cars and commercial vehicles.

The future of the Korean automobile industry is indeed bright, and there seems to be no obvious reason for it to decline. On the contrary, it will most probably grow and thereby provide the steel industry in Korea with a very significant future market. Korea has grown significantly in

Table 3.4 Locations and Capacities of Korean Auto Plants

Company	*Location*	*Estimated Annual Capacity*
Asia	Kwangju	222,000
Daewoo	Pupyong	500,000
	Pusan	6,000
	Kunsan	300,000
Daewoo Heavy Industries	Changwon	240,000
Halla	Umsong	1,000
Hyundai	Inju	300,000
	Chonju	100,000
	Ulsan	1,260,000
Kia	Asan	650,000
	Sohan	350,000
Samsung Heavy Industries	Changwon	5,000
	Taegu	200,000
Samsung Motors	Pusan	120,000
Ssangyong	Ulsan	100,000

Source: *Ward's World Auto Atlas and Directory*, 1998 Edition, p. 172.

terms of automobile production and usage since 1982. In that year, total vehicle production was 221 thousand, while registrations were 785 thousand, with a population of 44.1 million. The number of persons per car was 108. By 1997, total production was 2.6 million and total registrations were 9.2 million, with a population of 45.5 million. Although the number of persons per car had declined to six, this was still significantly above the 2.2-person level prevailing in Europe and North America, so that the Korean auto market by no means is near a saturation point. This leaves considerable room for more automotive expansion.

The Korean auto industry's growth in terms of production and exports continues to be positive. As previously indicated, production figures are expected to grow in the next 20 years, so the industry should provide a significant and growing market for steel over that period. A recent capital program completed by POSCO places the company in a favorable position to meet the demands of the auto industry, which have become increasingly more exacting in terms of the quality of the steel required. There has been a significant increase in demand for galvanized steel, insofar as it is corrosion resistant. POSCO has responded with additional galvanizing capacity, as well as the company's recently installed No. 4 six-high cold-reduction mill at Kwangyang, which delivers cold-rolled sheets of the highest quality.

Figure 3.1 shows one of POSCO's six-high cold-reduction mills, which can produce higher quality steel by virtue of the additional pressure applied as the steel is being reduced in thickness. In addition, adjustments made to the hot-strip mill, such as side shifting and roll bending, have also served to improve the quality of steel sheets. Figure 3.2 shows the finishing stands of a hot-strip mill at POSCO. The company has made these adjustments and additions, so that it will be able to meet any of the product requirements of the automotive industry.

One unfortunate development in respect to the Korean auto industry has been the difficulty surrounding the possible sale of financially troubled Daewoo. One prospective buyer, the Ford Motor Company, has withdrawn its bid, although Daewoo and General Motors have since been holding talks that could ultimately lead to GM purchasing a controlling stake. There also are other possible bidders, and, in time, the situation should reach a resolution.

In the past few years, the aluminum and plastics industries have at-

Figure 3.1 POSCO Six-High Cold-Reduction Mill

Figure 3.2 Finishing Stands of a POSCO Hot-Strip Mill

tempted to penetrate the automobile market and have had a certain amount of success. Table 3.5 indicates that in 1977, aluminum accounted for 2.6 percent of the average car's weight and plastics 4.6 percent, much less than the 74.8 percent accounted for by steel. Currently, steel's percentage is 65.1 percent, while aluminum and plastics have increased their shares to 7.5 percent and 7.6 percent, respectively.

There will be competition in the years ahead among these materials, and as a consequence, steel companies, including POSCO, have striven to improve the products and services they provide to the auto industry. POSCO's active participation in the international Ultra-Light Steel Auto Body (ULSAB) project is one example of the initiatives under way that seek to forestall the substitution of aluminum and plastics for auto-body use, thereby ensuring that the auto industry, including that in Korea, continues to represent one of the leading markets for steel well into the future.

APPLIANCES

The term "appliances" covers many items. The manufacture of some appliances consumes significant amounts of steel, while others such as television sets primarily use quantities of plastics and glass. In terms of steel consumption, the refrigerator uses more in its construction than many other items, including washing machines and microwave ovens. In Korea, appliance manufacturing has grown tremendously in the last decade. In

Table 3.5 Materials in a Typical Family Vehicle: Selected Years 1977–2000

	Iron & Steel		*Plastics*		*Aluminum*		*Other**		*Total*
	lbs.	*% wt.*	*lbs.*	*% wt.*	*lbs.*	*% wt.*	*lbs.*	*% wt.*	*Wt./lbs.*
1977	2,742	74.8	168	4.6	97	2.6	659	18.0	3,666
1987	2,235	70.3	222	7.0	146	4.6	575	18.1	3,178
1992	2,141	68.3	243	7.7	174	5.6	578	18.4	3,136
1998	2,174	66.6	244	7.5	224	6.9	620	19.0	3,262
1999	2,161	66.0	245	7.5	235	7.2	633	19.3	3,274
2000	2,140	65.1	249	7.6	246	7.5	651	19.8	3,286

*"Other" includes copper and brass, powder metals, zinc, magnesium, fluids, lubricants, rubber, glass, and miscellaneous materials.

Source: *American Metal Market* industry reports compiled in *Wards Automotive Yearbook*, 2000, 62nd edition, p. 47.

1990, as Table 3.6 shows, the total figure for all appliances, which includes refrigerators, washing machines, microwave ovens, air conditioners, VCRs, and color televisions, was 32.9 million units, which by 1995 had grown to 49.1 million units. In 1996, a record was established with the production of 52.2 million units, and in the next two years, this was trimmed by the Asian economic crisis to 46.4 million units in 1997 and to 38.2 million units in 1998. A significant upturn occurred in 1999, when production increased to 47.5 million units.

The renewed strength in the appliance market since 1999 can be traced to a recovery in consumer purchasing power, as well as a reduction in prices due to the abolition of the appliance tax, which allowed prices to be reduced by 12 percent. Further, there have been sales promotion activities on the part of producers, resulting in an increase in sales. Exports of most appliances have also increased due to a recovery in the economies of Southeast Asia.

A number of appliances are considered home appliances and have become necessities of modern living, particularly when household income grows in prosperous years. In 1999, Korean refrigerator production totaled 4.7 million units. This consumed a significant amount of steel, particularly in relation to the other appliances, such as microwave ovens, which consumed relatively small amounts for manufacturing each individual unit.

There are a number of Korean companies that have prospered in the appliance business. Daewoo, which currently has financial problems with its automobile business, has been Korea's number one exporter of five major home appliances: color televisions, VCRs, refrigerators, washing machines, and microwave ovens. In 1997, their combined export volume was 23.1 percent above the 1996 figure. Other companies also have registered favorable results in the manufacture and sale of appliances. Samsung Electronics set an export goal for 1999 of 1.2 million appliance units and sent 400 million won worth of appliances to China in that year's first quarter alone, which was about the same amount as it sold in all of 1996. Another appliance maker, LG Electronics, has become one of the world's largest exporters of air conditioners behind Matsushita and Mitsubishi, and a surge in export sales starting in 1999 has made air conditioners one of Korea's leading home appliance exports.[4]

During 1997–99, the Korean steel industry sold large quantities of its

Table 3.6 Appliances Manufactured in Korea: 1990–99 (thousands of units)

	1990	*1991*	*1992*	*1993*	*1994*	*1995*	*1996*	*1997*	*1998*	*1999*
Refrigerators	2,827	3,500	3,296	3,585	3,943	3,975	4,292	4,256	3,790	4,734
Washing Machines	2,163	2,157	1,895	2,199	2,443	2,827	2,878	2,967	2,643	2,822
Microwave Ovens	6,061	7,174	7,072	8,279	10,209	10,487	10,099	10,989	9,586	9,762
Air Conditioners	655	836	826	658	850	1,487	1,929	1,955	2,337	3,839
Color TVs	12,893	13,449	14,992	15,375	16,999	18,555	21,469	16,428	12,763	15,556
VCRs	8,305	9,336	9,532	10,416	11,785	11,792	11,560	9,781	7,070	10,804
TOTAL	32,904	36,452	37,613	40,512	46,229	49,123	52,227	46,376	38,189	47,517

Source: Pohang Iron and Steel Company, Ltd.

products to the appliance industry. In 1997, it sold 718 thousand tonnes, which unfortunately declined in 1998, a year of financial crisis, to 643 thousand tonnes, but recovered to 831 thousand tonnes in 1999. Conditions look good for the continued growth of the South Korean appliance industry, both on the domestic front and in the export market. Consequently, significant amounts of steel should be used by appliance manufacturers, and much of this will be in the form of sheets.

The attitude of the manufacturers of appliances toward the future of their business can be seen in the fact that the three major producers have decided to increase their production of refrigerators to become larger factors in the world market. They are improving the quality of their products and, recently, have produced refrigerators that have reduced the use of electricity 30 to 50 percent, compared to the power needs of many present-day models. This has increased sales considerably both in the domestic and export markets.

In the Korean manufacture of boilers, there has also been a dramatic change. Boilers were operated with oil. However, there has been a major shift from oil to gas in the last year or two, and many of the oil-operated boilers have been replaced by gas. One of the companies manufacturing a large number of gas boilers has exceeded one million units in sales. In fact, the company has had a very active selling campaign for five years in a row, and this will probably continue for the next year. The manufacture of an increasing volume of gas-fired boilers will require a significant amount of steel, as will the more efficient refrigerators that are being produced to replace those using large amounts of electricity.[5]

Appliances will continue to be produced in large numbers in South Korea and, thus, it will be a market that requires significant amounts of steel in the years ahead. This is particularly true of boilers, which manufacturers anticipate will see demand growth of 10 to 15 percent annually. As for refrigerators, which are one of the largest selling appliances in the group, 4.3 million were sold in 1997 in the domestic and export markets. One of the largest producers, Samsung, has improved the products being offered, not only domestically, but also in the export market. Samsung has stated that it will upscale the refrigerators being sent to the United States, both in terms of size and their efficiency in using electric power. The newest refrigerators being brought on the market are much larger, with a capacity of 680 liters, and have brushless motor circuits.

Generally speaking, the demand for home appliances, particularly refrigerators, is for the higher-priced models. This was revived in 1999, one year after the financial crisis. In fact, LG Electronics sold 2,500 two-door refrigerators in less than a month after they were first offered to the public, even though these refrigerators sell for about three times as much as their smaller counterparts. There is generally a stronger demand for the higher-priced appliances than for the lower-priced units.

The efforts placed on product improvement by Korean manufacturers of consumer durable goods resulted in a significant increase in the sales of these items in 1999. Durable goods, such as home appliances and automobiles, increased in sales to 15.9 trillion won. This is a 41 percent increase from 11.4 trillion won recorded in 1998. This increase is attributed to the fact that purchases of these items were postponed in 1998 due to the financial crisis. The principal item responsible for the increase of 41 percent in sales was the automobile, sales of which increased by over 80 percent to 4.6 trillion won from 2.5 trillion in 1998. Among electric household appliances, VCR sales rose 32.7 percent, television sales 27.3 percent, and refrigerator sales 20.5 percent.[6]

In terms of the growth of appliance demand, it would seem that steady increases are likely. Consequently, the demand for steel to be consumed by appliance manufacturers in the years ahead will grow with increases in Korea's population and, what is more significant, increases in the number of Korean households.

SHIPBUILDING

The nature of the shipbuilding industry in Korea has changed dramatically. During the last 30 years, a number of shipbuilding companies have changed their scope of activity from the construction of small coastal ships for domestic owners to large oceangoing vessels for export; and today they are capable of constructing both bulk carriers and tankers up to about 300 thousand deadweight tons (DWT).

The Korean shipbuilding industry started to build ships for export in the 1970s and gained a 6.3 percent worldwide share for large vessels. Currently, Korean shipyards have approximately 40 percent of the world order book. This large share of ships on order reflects new shipbuilding technology in a number of yards and a work force trained to use this technology effectively. Improvements in productivity and labor skills coupled with competitive pricing have resulted in a large number of orders placed with Korean yards by shipowners around the world.

A number of new yards were constructed in the 1970s, several of which were built by the heavy-industries companies. In 1972, Hyundai Heavy Industries Ltd. broke ground for what is now considered to be the number one shipyard in the world. The shipyard is located in Ulsan, on the southeastern tip of the Korean peninsula. Its progress in shipbuilding can be seen by the fact that in its first 27 years of operation the yard produced 780 ships in a large variety of classes totaling 56 million DWT, delivered to 167 owners from 37 different nations. In some years, Hyundai Heavy Industries has been the leading ship producer. Not only does the yard operate with highly skilled workers and with the most modern shipbuilding technology, it is also unique in having two research divisions to further enhance its technology.

The Samho yard was built by the Halla Group in 1995 and is currently under Hyundai's management after Halla's financial collapse. This yard is capable of building up to 1.5 million gross tons of vessels a year of a large range of vessel types, including very large cargo carriers (VLCCs), container ships, LNP and LPG carriers, and other specialized vessels, as well as offshore structures.

Daewoo Heavy Industries Ltd. built another large shipyard in 1978. This facility has passed through several changes and has become one of the leading companies in world shipbuilding. The yard builds vessels with a variety of the most modern technologies and has constructed both offshore and onshore facilities and vessels ranging from naval ships to ferries. As of April 1999, the yard had delivered more than 300 commercial ships and 47 naval vessels. It currently is a leading builder of sophisticated, large LNG carriers with more than a dozen vessels on order.

Samsung Heavy Industries Co. Ltd. entered the shipbuilding business on Koje Island in 1977. By the end of 1998, the company had delivered more than 200 ships and secured a backlog of 76 vessels amounting to five million gross tons. The shipyard has very sophisticated facilities capable of producing 1.8 million gross tons per year.

Hanjin Heavy Industries Co. Ltd. was a pioneer in creating the Korean shipbuilding industry. Established in 1937, the shipyard has built various kinds of vessels such as gas carriers and a cable-laying repair vessel.

Another significant Korean shipyard is Daedong Shipbuilding Co. Ltd., which specializes in small and medium-size vessels up to Panamax beam. The company has produced more than 400 vessels in its Pusan shipyard since 1962. In addition to the above-mentioned shipyards, Daesun Shipbuilding and Marine Engineering Co. Ltd. has built hundreds of smaller vessels.

Since the 1970s, the Korean shipbuilding industry has achieved a prominent international position. There seems to be no question that the Korean shipyards will maintain their high rate of new-building construction and will remain an important market for steel.

In the three-year period from 1997 to 1999, POSCO shipped sizable tonnages to the shipbuilding industry. In 1997, it shipped 792 thousand tonnes, increasing by almost 50 percent to 1.15 million tonnes in 1998. In 1999, tonnage shipped increased further to 1.29 million tonnes or about 7.8 percent of POSCO's total shipments.

Barring unforeseen economic changes in Korean shipbuilding economics, the biggest threat to the future of the yards is the expansion of shipbuilding capacity in neighboring China. However, it will take time before China will have acquired the technology in all of its yards to compete against the Korean yards for other than standard bulk carriers and tankers.

One item favorable for the shipbuilding industry is the existence of more than 1,000 single-hull oil tankers currently in operation. The International Maritime Organization has proposed the replacement of all single-hull tankers by double-hull tankers by 2015–2017. Enabling legislation is expected to be passed by this summer. If this takes place, there will be a tremendous flood of orders from shipowners the world over, and the Korean yards will benefit accordingly. Even if the current 2025 limitation on single-hull tankers is maintained, their replacement will still call for a large number of ships to be built, creating orders for both the Korean and Chinese shipbuilders.

RAILROADS

In Korea there is and has been a considerable amount of railroad activity. Currently, the focus of attention is a huge project known as the high-speed railroad, which when finished will connect Seoul with Pusan, a distance of 412 kilometers. It is expected that the road will be completely finished in 2010 at a cost of $16.3 billion. Individual sections will be completed and put into operation prior to 2010. The work is under the supervision of a French corporation, according to the model of a French train known as the Très Grande Vitesse. Part of the road will be electrified, and the rest will be operated by extremely powerful locomotives. Some of the rolling stock, including the locomotives, will be built abroad, as will some passenger cars.

POSCO does not produce rails, and thus it will not be involved in this aspect of the project. However, there are many other uses for steel, particularly structural steel, in the construction of railroad stations and facilities along the right of way. Since the project will not be completed until 2010, demand for significant amounts of steel will be generated before that time.

Another project, which is extremely ambitious, is now in the discussion stage. This involves a tunnel connecting Japan and Korea, permitting train traffic between the two countries. The train will then be routed across Korea to the Chinese border, connect there with the Siberian railroad, and finally terminate in Europe. Discussions so far have indicated that financing the project will be a huge undertaking. However, should the tunnel materialize, it will provide a very significant market for structural steel, plates, and other steel products.

TINPLATE

Tinplate is used for a variety of applications, principally for the manufacture of cans for food and beverages. These containers are produced in a large variety of sizes and shapes and number in the millions in terms of volume of cans made. In Korea, in addition to POSCO, there are three other manufacturers of tinplate. They are Dong Bu, Tong Yang, and Shin Wha. The total capacity of the industry is 935 thousand tonnes, with POSCO operating 180 thousand tonnes, Dong Bu 380 thousand tonnes, Tong Yang 275 thousand, and Shin Wha 100 thousand tonnes.

The total volume of tinplate sold to users in 1997 was 599 thousand tonnes. This included the domestic market and exports. In 1997, the total domestic market was 388 thousand tonnes, while exports were 211 thousand tonnes. In 1998, a poor year generally throughout Asia, Korea's total domestic market was 297 thousand tonnes, while exports rose to 380 thousand tonnes. In 1999, the Korean domestic market rose to 386 thousand tonnes, while exports were at 388 thousand tonnes. For 2000, the estimate is 389 thousand tonnes for the domestic market and 442 thousand for export.

In view of activity in the last three years, which has remained steady in the domestic market and has increased significantly in the export market, the future demand for tinplate in the next few years should be steady and possibly even rise to some extent. Currently, the consumers of tinplate in the domestic market are Hanil Can (17 percent), Seung Il Can (13 percent), Dae Ryuk Can (8 percent), Sam Hwa Can (7 percent), and all others 55 percent.[7]

OIL AND GAS

Korea has no reserves of oil or gas and, consequently, must import both of these products, which are essential to the nation's economic development. In 1998, imported oil and liquefied gas were used to produce 60 percent of the total energy consumed in the country. This required oil consumption of two million bbl-d. The oil is refined into usable products in five major refineries that have a total capacity in excess of 2.5 million bbl-d.

The Korean National Oil Corporation keeps a 60-day supply of oil on hand in the event that normal imports are interrupted for any reason. The company is also searching for oil in a number of countries outside of Korea. These include not only countries in South America and the Middle East, but also include drilling offshore in the North Sea. All of this is in the hope that significant deposits of oil may be found that will make Korea more energy independent. The exploration requires drilling equipment, which is made of steel. This is particularly true of the various types of drilling and line pipe for the production and transportation of petroleum, oils, and lubricants. As the Korean National Oil Corporation continues its exploration and development, looking for reserves of oil in various places, it will consume significant amounts of steel, and this should result in a continuing demand. Currently, Korea is the fourth-largest importer of oil of all the nations in the world and the second-largest importer of liquefied natural gas.

Korea will have to construct offshore-drilling platforms, which will require considerable steel. These will be provided by the shipbuilding industry, which has been building offshore platforms for several years. Thus, the steel industry can count on a significant amount of steel demand from the oil and gas industry, as it continues to expand in order to relieve South Korea of its complete dependence on oil and gas imports. In terms of refining the oil into various products, there are five refineries that are listed as major operations. They include Ulsan (817 thousand bbl-d), Onsan (500 thousand bbl-d), Yocheon (634 thousand bbl-d), Daesan (310 thousand bbl-d), and Inchon (270 thousand bbl-d).

In order to facilitate additional handling of LNG imports, there has been a joint venture agreed to by POSCO, M.W. Kellogg Co. of the United States, and Mitsubishi Corporation of Japan to build an LNG terminal to convert 1.2 million tonnes a year on a site next to POSCO's Kwangyang steel mill. POSCO will receive LNG to fire its house-power generators. This is scheduled to be completed in 2002 at an estimated cost of $400 million.[8]

PIPE AND TUBING

Among the steel-consuming industries which POSCO serves, several stand out in terms of the steel demand they generate. Among these are automobiles, service centers, rerollers, and a category referred to as "pipe." POSCO does not make pipe, but provides considerable amounts of steel to other companies in Korea that do make various types of pipe, their aggregate capacity approaching 8.5 million tonnes.

Two of the largest pipe manufacturers in Korea are Hyundai Steel, located in Ulsan with a little less than one million tonnes of capacity, and SeAH, located in Pohang with a capacity in excess of 1.1 million tonnes. Hyundai Pipe makes electric-resistance weld pipe, spiral weld (SW) pipe, and submerged-arc weld (SAW) pipe, and SeAH Steel also makes electric-resistance weld pipe, SW pipe, and SAW pipe, all of which have multiple uses. Other companies that make significant tonnes of pipe include

Shinho Steel; MI Ju Steel; Korea Cast Iron Pipe; Dongbu Steel; and Changwon Specialty Steel, which makes seamless pipe.

Among the various categories of pipe, either welded or seamless, are standard pipe, pressure pipe, pressure tubes, line pipe, oil-country tubular goods, rigid conduit, mechanical tubing, and structural tubing. Each of these categories has different uses which can be described as follows:

1. Standard pipe is used in plumbing and heating for homes, office buildings, and factories; in water supply systems; fire prevention systems; irrigation; refrigeration; transportation; mining; snow melting; and for general structural purposes.
2. Pressure pipe is used in the steam and hydraulic piping systems of plants generating electric power, in plants refining petroleum, and in many similar industrial uses.
3. Pressure tubes are used to carry fluids at high temperatures or pressures or both and for applications involving exposure to external heat.
4. Line pipe is used for conveying gas, oil, and liquefied petroleum products, or water. It is also used to transport solid materials, such as very finely ground coal.
5. Oil-country tubular goods consist of three types of pipe used in wells, namely, casing, drill pipe, and tubing.
6. Rigid conduit is used for the protection of wiring systems.
7. Mechanical tubing is used, as its name implies, for mechanical purposes.
8. Structural tubing is used for the welded, riveted, or bolted construction of bridge for buildings, and for general structural jobs.

Every year, POSCO has shipped significant amounts of steel to the pipe industry. In 1998, which was a depressed year, the tonnage was 2.9 million tonnes or 16.7 percent of POSCO's total domestic shipments for that year. In 1999, shipments to the pipe industry amounted to 3.2 million tonnes or 16.6 percent of the company's total domestic shipments for that year. The tonnage in 2000 was 3.1 million tonnes or 15.4 percent of total domestic shipments. Pipe has such a variety of uses that it is expected to remain a steady consumer of steel for decades to come.

CONSTRUCTION

The construction industry covers a multitude of projects including infrastructure building, the construction of tall buildings, residential construction, and a number of other types of building activity. POSCO's steel shipments from 1997 through 1999 to what is termed the construction industry have been in excess of one million tonnes each year. They were 1.3 million tonnes in 1997, or 8 percent of POSCO's shipments. In 1998, they dropped slightly to 1.2 million tonnes, or a 7.8 percent share. In 1999, the figure was just slightly above one million tonnes, which was 6.3 percent of POSCO's shipments.

The construction industry consumed much more than this tonnage, since it relies heavily on Korea's service centers to supply a considerable tonnage of its steel requirements. Recently, steel has been used quite liberally in residential construction, and in many cases, it is used for roofing and interior applications, such as doors and window frames. It is used in many items in infrastructure, such as highways and bridges. As a consumer of steel, the construction industry is one of the largest customers of the service centers. In the United States, for example, the construction industry accounts for 28 percent of the total sales of the service centers.

One type of infrastructure building that consumes substantial quantities of steel is bridge construction, which has been very active in Korea during recent years (See Table 3.7). During 1993–99, an average of 1,039 contracts were placed each year for the construction of general bridges and railway bridges combined, creating a demand for plates, structurals, decking, wire rope, reinforcing bar, and other types of steel.

Table 3.7 Contracts for Bridge Construction in Korea: 1993–1999

	General Bridges		*Railway Bridges*	
Year	*Number of Contracts*	*in Billion Won*	*Number of Contracts*	*in Billion Won*
1993	569	21.8	26	5.0
1994	702	42.4	17	2.4
1997	1,248	164.7	35	10.9
1998	1,325	159.5	44	15.0
1999	1,176	110.4	53	14.8

Source: Pohang Iron and Steel Company, Ltd.

To increase the use of galvanized steel coils in the manufacture of framing components for residential buildings, core structural technologies have been developed with the aid of POSCO's Research Institute of Science and Technology (RIST), and government support has been sought for standards and specifications for steel-framing products. In promoting galvanized steel as the preferred material for residential construction, POSCO has participated in exhibitions, seminars, and education programs conducted under the auspices of the Steel House Club of the Korean Iron and Steel Association or KOSA.

SERVICE CENTERS

Steel companies usually sell and ship their products in large quantities. It is not uncommon to have a significant number of tonnes in one shipment to the automobile or shipbuilding industry. There are, however, many steel consumers who purchase limited amounts of steel, often in quantities of less than one tonne. Serving these consumers is the function of the steel service centers, which provide small amounts of steel to customers desiring limited quantities. This is in contrast to the steel mills, which usually provide steel in large quantities and, thus, find it convenient to have the service centers distribute the small amounts. Service centers, or as they are often called, distributors, can be looked on as the retail arm of the steel industry, which, in aggregate, sell millions of tonnes of steel products, although individual orders are negotiated and delivered in small amounts.

In terms of POSCO's shipments to the distributors or service centers, they have been large in total-delivery terms in recent years. In 1997, POSCO shipped 3.4 million tonnes to service centers. In 1998, the figure declined somewhat to 2.7 million tonnes because of the financial crisis, but in 1999, it improved to 3.9 million tonnes, which was 15.2 percent of the total tonnage shipped by POSCO in that year. Thus, distributors or service centers represent a significant outlet for POSCO's steel.

Another indication of their significance is the fact that in 1997, POSCO shipped 1.5 million tonnes of hot-rolled coils to 11 domestic service centers. Further, in 2000, 2.8 million tonnes of cold-rolled coils were shipped to 18 service centers. In terms of hot-rolled coils, among the large service-center buyers were Hanil Iron and Steel Company and Moon Bae Steel Company. In terms of cold-rolled coils, Saewoon Steel Company and Daechang Steel Company, Ltd. were significant buyers, with hundreds of thousands of tonnes purchased.

INDUSTRIAL INTERDEPENDENCE

In looking to the future, the focus of POSCO's market strategy is to maintain the company's long-standing commitment to maximizing customer satisfaction, on the one hand, and to adding value by continuously upgrading the product mix of the steel supplied to the steel-consuming industries on the other. Mainly because of POSCO's growth and development, Korea's self-sufficiency in supplying domestic steel consumers has improved markedly from just 20 percent in the early 1970's, before POSCO melted its first heat of steel, to more than 90 percent today. Born of industrial interdependence, the close working relationships that have been forged between POSCO and Korea's steel consumers have been invaluable to their interrelated achievements and, in turn, to the development of the entire Korean economy.

NOTES

1. Pohang Iron and Steel Company, Ltd., *Annual Report*, 1986, p. 13.
2. Pohang Iron and Steel Company, Ltd., *Annual Report*, 1997, p. 11.
3. *Ward's Automotive Reports*, January 8, 2001, p. 1.
4. *The Korea Herald*, May 26, 1999, p. 5.
5. *Business Korea*, November 1998, p. 67.
6. *The Korea Herald*, April 10, 2000, p. 5.
7. Hanil Can Company, *Welcome to Hanil Can* (Seoul: Hanil Can Company, 1998), p. 7.
8. U.S. Department of Energy, Energy Information Administration, *Country Analysis Briefs: South Korea*, September, 2000, p. 1.

4

The Technology Strategy

Over the last 30 years, POSCO has derived considerable benefit from the relative ease of international technology transfer in steel and from extensive development of the company's own technology base. Following the initial stage of the Pohang plant's construction, which relied on production facilities and vital financial support from Japan, POSCO's facility purchases have been made from a widely diversified base of the world's leading equipment suppliers, thereby affording the Pohang and Kwangyang plants access to the latest steel technologies the world has to offer. This practice of obtaining technological know-how from others continues to represent a key part of the company's technology strategy. The other part has been to pursue internal technology development as an essential requirement of achieving and maintaining a leadership position among the world's foremost steel-producing companies.

POSCO's technology strategy makes major commitments of human and financial capital to research and development, education, and technical training, and thereby seeks to foster in-house innovation, not only to drive improvements in the company's operating performance and competitive position, but also to contribute to educational and technical progress throughout Korea. In major initiatives to serve these objectives, POSCO has established two affiliates that rank among the world's leading research-oriented organizations, namely Pohang University of Science and Technology (POSTECH) and the Research Institute of Industrial Science and Technology (RIST). Both organizations and the company's own research division cooperate on joint research and development programs.

An overview of the physical facilities of POSTECH and RIST is shown in Figure 4.1.

POSTECH, a private technical university, was established on December 3, 1986 as Korea's first research-oriented institution of higher learning. In endeavoring to educate the nation's future scientific leaders, the university has earned an excellent worldwide reputation for its respected international faculty, advanced educational and research facilities, and maintenance of the highest in educational standards on both the undergraduate and graduate levels. The university has a current enrollment of 1,332 students at the undergraduate level and 1,496 taking graduate courses. The core curriculum encompasses physics, chemistry, mathematics, life science, materials science, mechanical engineering, industrial engineering, electric and electronic engineering, computer science, and chemical engineering. POSCO has long regarded POSTECH as affording the company one of the most desirable ways of returning corporate profits to Korean society.

RIST can trace its origins to POSCO's 1977 founding of the Technology Research Institute, which was reorganized to establish RIST on March 3, 1987. RIST has had a twofold purpose, namely to help develop new technologies of use to POSCO, and also to help Korea attain a greater self-reliance in science and technology. At present, RIST is staffed with 247 researchers, 191 engineers, and 40 administrators, while POSCO's own research division has 208 researchers, 187 engineers, and 40 administrators.[1]

In a typical division of research and development responsibilities, POSTECH pursues basic scientific research, RIST directs applied-technology development, and POSCO makes actual on-site applications of new technologies at both pilot plants and commercial production facilities. By means of this tripartite research and development strategy, POSCO has, on average, made more than 350 technological advances each of the last three years, ranging from the development of new steels and new steel coatings to process innovations such as ironmaking and rolling and environmental management. In recent years, smelting reduction and strip casting have been the focus of major research and development initiatives. In both of these areas, the commercialization of advanced process technologies would provide major payback in advancing the company's objective of developing "21st-century" approaches to steel opera-

Figure 4.1 Overview of POSTECH and RIST

tions that address economic, environmental, and energy issues of concern to the company and to steelmakers around the world.

SMELTING REDUCTION

Although POSCO has continuously modernized its ten blast furnaces and conceivably could have supplied all of its iron requirements solely through blast-furnace smelting, recognition of the economic and environmental advantages to be gained by circumventing the blast furnace's dependency on metallurgical coke led the company to build a COREX ironmaking plant, which went into operation at Pohang in November 1995. COREX was chosen as the only non-coke-smelting alternative to the blast furnace to have reached commercial status, a facility with a capacity to produce 300 thousand annual tonnes of molten iron having been first commercialized by South Africa's Iscor Ltd. starting in 1988.

The primary goal in developing COREX was to bypass the coking stage of the blast-furnace route and eliminate the need for coke by directly using a wide range of coals. The process was developed in Germany by Korf Engineering as the KR process for "kohle reduction," or coal reduction. Korf completed construction of a pilot plant at Kehl-Rhein in 1981, and thereafter a considerable tonnage of molten iron was produced from a variety of lump iron ores and pellets using a number of different coals, ranging from high-volatile bituminous to anthracite. The process was subsequently acquired by Voest-Alpine of Austria, renamed COREX, and scaled up to commercial proportions. In 1985, Iscor decided to install the first commercial COREX plant at its Pretoria Works to supplement the iron output of the blast furnace there.

The COREX plant built at Pohang represented a major advance in the technology's evolution by more than doubling the scale of the Iscor facility to attain an initial ironmaking capacity of 1,800 tonnes per day, the equivalent of 650 thousand annual tonnes, which, since the plant's startup, has been upgraded through process improvements to 700 thousand tonnes. Along with the COREX smelter, which produces a very large volume of gas, a power house was constructed to convert this off-gas into electricity. The total investment required was close to $400 million, and the molten iron produced gave the Pohang plant the capability to produce an additional one million annual tonnes of crude steel. The COREX plant at Pohang is shown in Figure 4.2.

POSCO's objectives in installing its COREX plant extended beyond the production of supplemental iron to further support Pohang's steelmaking capabilities. The COREX plant also gave the company the equivalent of a commercially run research laboratory for day-to-day use in continuing to advance the development of cokeless smelting reduction.

In deciding to build a COREX plant, POSCO believed that as the leading smelting-reduction alternative to the blast furnace, the process was a step in the right direction. However, the company also recognized that there was room for improvement, particularly because the process was capable of using only lumpy ore or pellets with a diameter of 8 to 30mm. Accordingly, POSCO realized the importance of advancing the process through the development of fine-ore technology. Toward this end, the company in 1990 had formed the "New Project 1 R&D Team," comprising 17 top researchers from RIST, and starting in 1992, the team worked in close cooperation with Voest-Alpine's researchers in jointly investigating the use of 100 percent fine ore in what ultimately would be a modification of COREX to be known as the FINEX process. The development of FINEX was supported by the Korean steel industry, through its Korea New Steel Technology Research Association, with POSCO playing a principal role.[2]

As a result of the research effort, a technology was developed to inject fine ores into the COREX smelter at up to 15 percent of the iron-ore charge, and 100 percent fine ore was successfully tested at a not-fluidized-bed model plant with a capacity to process 15 tonnes of fine ore per day. Once results using the model plant established the feasibility and economic viability of FINEX, development of the process was advanced on February 3, 1998, when POSCO and RIST broke ground at Pohang for the construction of a pilot plant with a daily capacity of 150 tonnes.[3]

Started up in July 1999, test operations at the FINEX pilot plant, in which Voest-Alpine also has been a participant, established that the use of less expensive fine ore without costly pre-processing affords reductions in both investment and net production costs, while also yielding substantial environmental benefit relative to conventional blast-furnace ironmaking with its need for sinter plants and coke ovens. In the FINEX process, fine ore is reduced to 80–90 percent metallic iron in a series of fluidized

Figure 4.2 COREX Ironmaking Plant at Pohang

beds, and the remaining iron oxides are reduced, carburized, and melted with the formation of a slag in a melter-gasifier, which is analogous to that employed in the COREX process. As coal is gasified with oxygen in the melter-gasifier, the resulting reduction gases are supplied to the fluidized-bed system, and the remaining export gas is similar in quality to that generated by COREX.[4]

Given the results of the pilot project, POSCO in December 2000 started construction of a FINEX demonstration plant designed to produce 600 thousand tonnes of molten iron annually. Scheduled for completion in February 2003, the plant is being built by POSCO, which is also providing the necessary equipment and project funding, while Voest-Alpine performs part of the project engineering. In the meantime, POSCO continues to produce COREX iron at Pohang and has exported the operation and maintenance technology developed there to Saldanha Steel of South Africa and Jindal Vijayanagar Steel of India, which also operate COREX ironmaking plants.

NEAR-NET-SHAPE CASTING

Near-net-shape casting is the other advanced process technology that has been a major focus of POSCO's research and development efforts. The technology encompasses methods by which liquid steel is continuously cast into shapes more nearly approximating, or even representing, the final dimensions to be achieved in the hot-finishing stage of production. In seeking to both improve and revolutionize the output of sheets and plates, POSCO's strategy in regard to near-net-shape casting has been twofold. First, the company has invested in a commercial minimill that employs thin-slab casting, thereby keeping apprised of and deriving technical and economic benefit from this lower-cost, streamlined alternative for producing many types of hot-rolled sheets. And second, for more than a decade, the company has conducted internal research and development into continuous strip casting, a breakthrough technology that seeks to completely eliminate the slabbing and hot-rolling stages of production.

The use of near-net-shape casting in the production of sheet steel has long been an elusive goal of steelmakers and inventors, including Sir Henry Bessemer, who around 1860 conceived of a dual-rolled continuous caster as a possible method for directly transforming liquid steel into finished plates. The ensuing years brought a long series of attempts using machines of varying design, all aimed at continuously casting useful shapes. However, more than a century would pass before meaningful progress was made, when Nucor Corporation in mid-1989 successfully started up the first commercial thin-slab casting machine at Crawfordsville, Indiana. The technology employed was based on the designs of Schloemann Siemag AG (SMS) in Germany, where pilot testing had been conducted starting in 1985. Marketed by SMS/Concast, Nucor had ordered the first commercial unit late in 1986.

Thin-slab casting seeks to minimize the plant and equipment needed to produce flat-rolled sheets and strip. In so doing, it lowers break-even plant capacity, and, so too, the capital and operating costs associated with conventional slab-casting and hot-strip-mill operations. In most conventional sheet plants, basic-oxygen (BOF) steel is continuously cast into slabs with approximate thicknesses of 200 to 300mm. In many cases, these "thick" slabs are then inventoried and require reheating prior to rolling. They are then passed through a conventional hot-strip mill's reversing or multiple roughing stands, numbering as many as six, and then, once a thickness of 25 to 38mm is achieved, are rolled on the mill's six- or seven-stand finishing train to a minimum thickness of 1.3mm in the finished hot-rolled coil. Nucor's original "thin" slabs were 87mm in thickness to begin with, eliminating the need for conventional reheating and roughing operations.

In January 1995, POSCO began construction at Kwangyang of a thin-slab minimill based on technology known as the ISP process for in-line strip production. The technology had evolved in 1987 from trials conducted at Mannesmann Steelworks in Duisburg, Germany and from Mannesmann's joint development work with Italy's Arvedi Group in converting a standard continuous caster to produce thin slabs at Acciaieria Tubificio Arvedi in Cremona, Italy. Subsequently marketed by Mannesmann-Demag, the first commercial ISP plant had been ordered by Arvedi in May 1989 and had started break-in operations toward the end of 1991.

POSCO initially contracted with Mannesmann-Demag in May 1994 for a minimill using ISP technology, modified to POSCO's specifications and capable of producing a million annual tonnes of hot-rolled coils. Then, six months later, POSCO ordered a second melt shop and second ISP thin-slab caster to feed additional slabs to the minimill's hot-strip finishing mill and thereby increase hot-rolled capacity to 1.8 million annual tonnes. POSCO Engineering and Construction Company, Ltd. or POSEC supplied all of the minimill's engineering, procurement, construction, and auxiliary requirements. POSEC, the engineering and construction arm of POSCO, had been established in December 1994 through the merger of Geoyang Development Company, Ltd. and POSCO Engineering Company, Ltd. Coincidentally, the groundbreaking for the new minimill was held during the same month.

The Kwangyang minimill's facilities are arranged for a continuous flow of operations from electric-furnace charging to melting, secondary refining, continuous thin-slab casting, reduction rolling, transfer-bar induction heating, coil handling, and finally, hot-strip rolling. These continuously linked operating steps are conducted within a production line having a total length of 196m, the resulting end product being coiled hot-rolled strip ranging in width from 900 to 1350mm, with a maximum coil weight of 24 tonnes.[5]

The minimill has two twin-shell, DC-type, electric-arc steelmaking furnaces, each turning out a 130-tonne heat of steel every 50 to 60 minutes and each followed in line by a ladle-refining furnace and a vacuum degasser. Equipped for charge preheating, the electric furnaces melt combinations of revert scrap, blast-furnace hot metal, pig iron, and hot-briquetted iron or HBI. Hot metal from Kwangyang's No. 5 blast furnace often represents some 40 percent of the total furnace charge, which, compared to the use of 100 percent scrap, shortens the tap-to-tap and power-on times by 13 percent and 18 percent, respectively, and electric-power consumption by 26 percent. The use of hot metal also upgrades the quality of the steel produced.

Molten steel is supplied to each thin-slab caster via a revolving two-turret system, which supports and moves covered, steel-filled ladles, as well as prepared tundishes into position for casting. Compared to conventional, rail-operated units, the turret system permits a faster, more flexible changing of ladles and tundishes during multiple-sequence casting and is more accurate in positioning the tundish's submerged nozzle into the casting machine's mold. In the casting process, a steel-filled tundish, its cover pressurized with argon gas, feeds 30 tonnes of liquid steel into the caster mold at a precise rate, which is metered by an eddy-current-equipped level-control system.

The 75mm-thick, straight caster mold is equipped with a hydraulically driven resonance-mold-oscillating system to ensure surface quality in the cast slab and permit safe operation at casting speeds averaging some 4.5m per minute. The casting speed is controlled by an electromagnetic brake, which also minimizes turbulence and promotes a uniform dispersal of the liquid steel inside the mold. The casting machine itself, which employs a vertical-bending design beyond the mold and has a metallurgical length of 10.7mm, cast rolls the formed slab with a still-liquid core, reducing the slab thickness to 67.5mm. Internal water cooling of the mold and other caster components occurs throughout the casting process, while controlled misting with air is used to cool the cast slab. After the slab is unbended in the caster's final segment, a hydraulic desealer is used to remove any surface scale and residual casting powder.

Next in line, each of the minimill's thin-slab casters is equipped with a two-stand reduction unit in which the slab's thickness is reduced from 67.5mm to 25–30mm. The operating speed of the unit's first stand corresponds with and is governed by the exit speed of the continuous caster. Both reduction stands are equipped with automatic gauge control and work-roll bending to insure the flatness of the final product. After reduction, a pendulum shear cuts the bar into lengths determined by the required weights of the finished coils to be rolled.

After exiting the reduction unit, cut slabs or transfer bars in need of being raised to a required rolling temperature pass through an induction-heating system that is 6.0mm in length, before entering a coil-handling system, comprising an acceleration furnace, two coiling stations, one uncoiler, and two holding furnaces for the buffer storage of up to five coiled transfer bars. The coil-handling system provides sufficient buffer capacity to accommodate differences between the rates at which transfer bars are produced and subsequently required for rolling into finished coils.

Transfer-bar output from Kwangwang minimill's two thin-slab casters is processed on a single hot-strip mill with five four-high stands. The mill is equipped with hydraulic gauge control, electrical loopers, roll bending on all five stands, and pair-cross rolling on the three middle stands. The mill's runout table is equipped with laminar-flow-cooling and is followed by two downcoilers and a coil-conveying system. In 2000, the Kwangyang minimill operated at 99.3 percent of capacity and achieved a full year's output of just under 1.8 million tonnes.

At the same time that thin-slab casting has become an integral part of day-to-day commercial operations at the Kwangyang plant, POSCO has continued its long-standing research and development effort to realize the full potential of near-net-shape casting by internally pursuing advances in continuous strip-casting technology. The company initiated its strip-casting program in 1989, when RIST and Davy McKee of the United Kingdom joined forces to develop a pilot strip caster that would be capable of directly shaping 350mm-wide steel plate. The joint effort focused its development work on integrating the machinery, materials, metallurgy, electronic controls, and automation needed to make the advanced process work, culminating in the construction of a pilot plant in September 1991. By the end of 1993, over 100 trial runs had been successfully carried out, and the hot coils produced were used by POSCO to make such test products as kitchenware and pipe.[6]

On January 27, 1994, ground was broken at Pohang for the construction of a large-scale continuous strip-casting pilot plant to directly produce 1300mm-wide plate as an initial step toward the technology's commercialization. The pilot plant was a cooperative effort among research teams from POSTECH and RIST, and engineering teams from Davy, Samsung Heavy Industries, and POSCO's advanced factory-automation subsidiary, POSCON. Completed in the first half of 1995, the pilot plant represented an initial investment of $17.6 million (See Figure 4.3).

The large-scale pilot plant can directly produce strip with a thickness of 2 to 6mm in just 0.2 seconds and has been successfully employed to continuously cast both carbon-steel and stainless-steel products. By eliminating the hot-rolling facilities required by steel plants with conventional continuous casters, POSCO's research and development in the field of continuous strip casting has the potential to ultimately deliver dramatic cost reduction and to scale down the length of strip-production facilities to just 60m, or one-tenth the 600 to 1,000m presently taken up by conventional flat-rolling plants.

The pilot-plant phase of POSCO's strip-casting program was completed early in 2001, and batch testing is currently under way. The results of this testing will be considered as part of an intensive review and analysis of all research and development findings, pilot-testing results, and other factors bearing on any decision about constructing a full-scale commercialized plant.

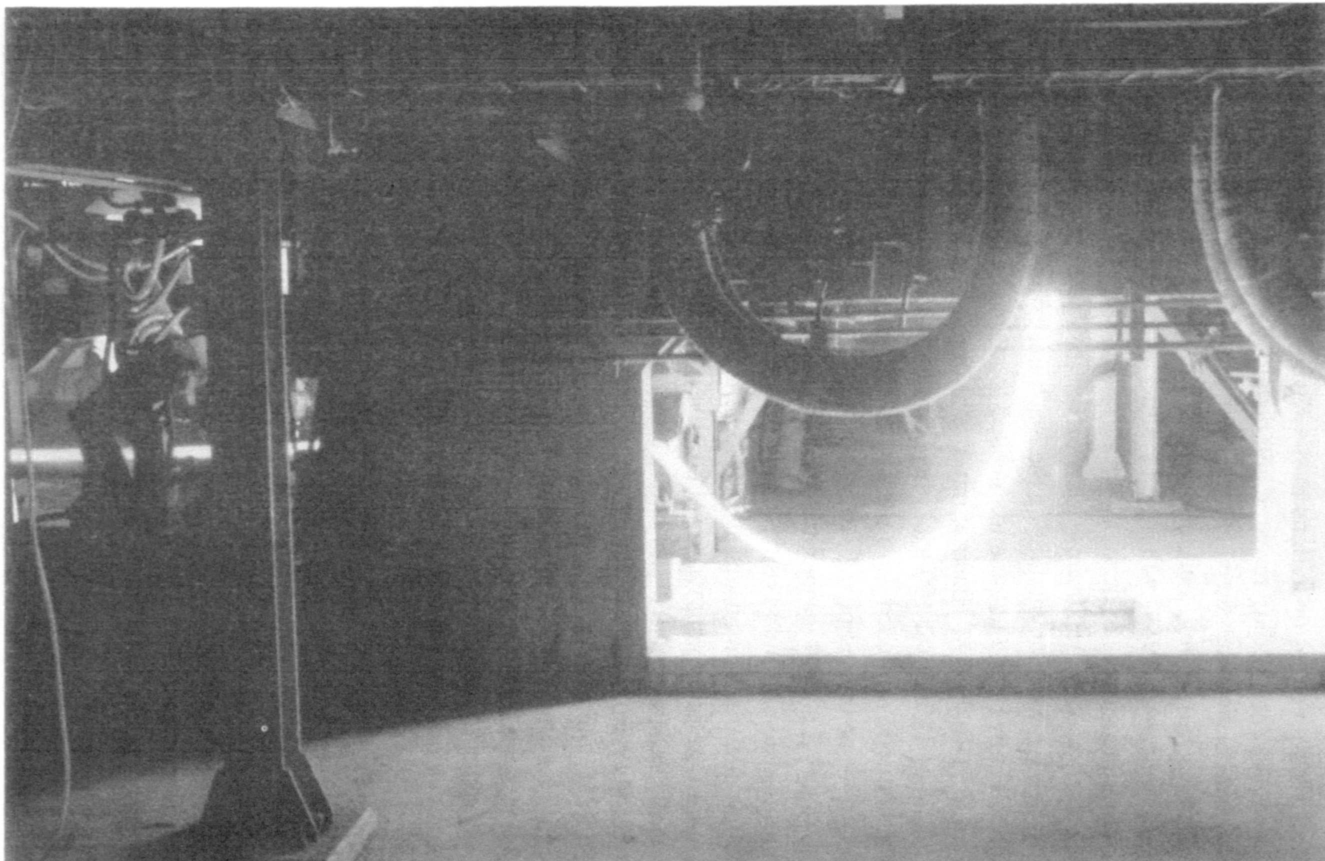

Figure 4.3 Strip Casting at POSCO

STEEL MILL OF THE 21ST CENTURY

The commercialization of smelting reduction and near-net-shape casting, more specifically strip casting, would represent dramatic progress toward fully realizing POSCO's vision of a 21st-century steel mill, a vision that has guided the design, facility selection, construction, and operating practices at the company's two integrated plants, particularly that at Kwangyang. Each of the plants was built to take advantage of the many prior advances in steel technology, all of which sought to increase the speed and efficiency and lower the cost of producing steels that had to meet ever-more-exacting quality standards, with the added objective of making the production process more compatible with environmental-protection and worker-safety requirements.

Historically, the major divisions or stages of steel-mill operations have comprised multiple batch-type production activities, resulting in numerous interruptions of their process flows and equally numerous instances of stocking in-process inventories. This batch-type approach placed limits on operating efficiency and involved heavy capital and environmental-cost burdens. It contributed to yield and energy losses, created conditions susceptible to the development of production bottlenecks, and restricted coordination between and among the major operating stages.

Most of the major advances in steel technology made in the last century were directed at overcoming the limits imposed by batch-type production. They did this by rationalizing process flows to permit an increasing degree of continuous operation. This technological emphasis was typified by continuous casting, continuous hot and cold rolling, and continuous annealing, all of which replaced less efficient batch-type operations.

The ideal steel plant permits a continuous flow of production from start to finish, which has been accomplished by Kwangyang's thin-slab minimill and its counterparts in other countries. The ideal has also been approached on an integrated production basis at Kwangyang, given the plant's straight-line layout and the direct, computerized linkage and coordination of its major stages of production. However, POSCO's technology strategy recognizes that beyond the accomplishments at Kwangyang, maintaining a competitive advantage will require the continuing adoption of future changes in iron and steel technology.

Both smelting reduction and continuous strip casting, which have been the focus of POSCO's research and development attention, eliminate costly steps in the production process, namely cokemaking and hot rolling, which gives the new technologies major potential to speed process flows in "2lst-century" steel plants. POSCO also continues to modernize and upgrade its computer information systems, which have been indispensable in coordinating functions within and between the major stages of production at both Pohang and Kwangyang. Combined with the implementation of new process technologies as they evolve, this is intended to keep both plants attuned to POSCO's vision for steel in the century ahead.

NOTES

1. Pohang Iron and Steel Company, Ltd., *Fact Book: POSCO 200l* (Seoul: Pohang Iron and Steel Company, Ltd., 2001), p. 28.
2. *POSCO Newsletter*, January 1994, p. 5.
3. Pohang Iron and Steel Company, Ltd., *Annual Report*, 1998, p. 8.
4. A. Zachetzsche, J.L. Schenk, A.F. Eberle, H.H. Lee, S.K. Jung, and I.O. Lee, "The Finex Fine-Ore Smelting Reduction Process," *Millenium Steel 2000* (London: Millenium Steel Publishing, 2000), pp. 75–80.
5. W.D. Huskonen, "POSCO Goes 'Mini' to Grow Larger," *Metalproducing* 33, July 1998, pp. 30–36. This article contains a full description of the Kwangyang minimill.
6. *POSCO Newsletter* February 1994, p. 3.

5

The Raw-Materials Strategy

Before there was a company called POSCO, Korea's burning desire to build an integrated steelworks was driven by a clear grasp of world economic history, which had demonstrated time and again that no nation can flourish and create a prosperous culture without iron. Through the first half of the last century, this also meant that industrial advancement mainly required that a nation be fortuitously endowed with its own or nearby deposits of the raw materials that ironmaking required. Access to iron ore, coking coal, and limestone, therefore, most often delineated the haves and have-nots among the nations of the world. Economists referred to this as the law of comparative advantage.

The law would have classified Korea as unquestionably disadvantaged, considering that the nation was virtually devoid of the raw materials it would take to effectively support integrated iron and steel production. Fortunately for Korea, by the late 1960s, when POSCO was founded, the law of comparative advantage had been all but nullified by a technological revolution in the worldwide transportation of bulk materials, including those required by integrated steel plants. The development of the large bulk cargo carrier had made it possible to ship raw materials across the oceans at low cost, so that large-scale integrated steel plants no longer had to be exclusively located in nations with indigenous raw-material resources.

The first development that had started to negate the law of comparative advantage occurred in the immediate post–World War II period, when it became apparent that iron-ore deposits in a number of the major steelmaking countries were quickly becoming depleted. This occasioned an urgent increase in exploration around the world, which by the end of the 1960s, resulted in the discovery of huge, rich ore deposits in Venezuela, Brazil, Australia, Canada, and Africa, thereby restoring a condition of plentiful supply, but forever changing the logistics of world steel production and diminishing the comparative advantage long held by traditional steel-producing nations.

At the same time as new ore deposits were being discovered, a new breed of oceangoing ships evolved to meet the demands of shipowners for carriers of increasing size. The sheer capacity of the new, larger carriers made them more economical to operate. In fact, a 47 percent savings in shipping costs could be realized by using a 100-thousand-DWT carrier rather than one of 18 thousand DWT. Like the new iron-ore discoveries that were being made, the advent of the larger carriers also diminished the comparative advantage once held by traditional steel producers, particularly those in the United States and Europe, where relatively shallow ports placed natural limits on the largest carriers that could be used.

Significantly, no such limits were imposed by the ports in Japan, which in time was to become the prototype emulated by POSCO in developing a raw-materials strategy that would be centered around the use of deepwater ports to accommodate very large bulk carriers.

Japanese steelmakers, despite having to contend with a severe lack of domestic raw materials for iron and steel production, increased their post–World War II crude steel output from 952 thousand tonnes in 1947 to 9.4 million tonnes in 1955 and to 93.2 million tonnes in 1970. To accomplish this, imports were relied upon to provide 90 percent of the iron ore and 60 to 70 percent of the coking coal needed to supply large blast furnaces.

The cost of acquiring these raw materials was greatly reduced by using large bulk cargo carriers, which made it possible to bring in single shiploads of iron ore and coking coal exceeding 100 thousand tonnes and 50 thousand tonnes, respectively. Raw-material costs were further reduced by building the steel plants on the seacoast at deepwater ports, which eliminated the need for rail haulage to inland locations. In this manner, Japanese steelmakers overcame a paucity of raw materials at home and had their pick of the finest iron ores and coking coals in the world, which considerably enhanced their blast-furnace operations.

DOMESTIC RESOURCES

When Korea launched its first five-year economic plan in 1962 and made the construction of an integrated steel plant a top national priority, several intensive surveys were conducted by teams of geologists and engineers to locate and determine the extent of the domestic resource base of ironmaking raw materials. Their findings confirmed that Korea lacked sufficient iron ore and coking coal to effectively support a significant blast-furnace operation. According to the Korean Geological Survey of 1966–67, there were iron-ore reserves of 80.3 million tonnes at 48 mines, including 15.1 million tonnes containing 40 percent iron and 65.2 million tonnes with an iron content of 20 to 40 percent. The ore consisted of magnetites, hematites, carbonates, and limonites, most of which were exceptionally high in silica.[1]

At the time, Japanese steelmakers were purchasing some 650 thousand annual tonnes of the 40 percent–iron, high-silica ore for blending with high-grade imported ores from other sources in order to increase the fluidity of their blast-furnace slag and thereby improve furnace operations. The use of an equivalent quantity of this ore in a Korean blast furnace would have depleted the higher-grade domestic reserves in less than 12 years. To be used in a blast furnace on an extensive basis, both the higher-grade ores and those having a lower iron content would have to be beneficiated at considerable cost.

Recognizing that the use of domestic iron ore would entail shortcomings, KISA, the international consortium that unsuccessfully attempted to build Korea's first integrated steel plant in the late 1960s, had recommended that supplemental supplies of high-grade ore be purchased from one or more international sources and through trans-shipment from Japan.[2] The consortium, however, did not foresee that Korea ever would have the large-scale integrated steel plants that POSCO would soon start to build, nor did the consortium envision that these plants would so effectively duplicate and improve upon the use of deepwater ports and large bulk carriers that were serving the Japanese steel industry so well.

As for coking coal suitable for metallurgical purposes, it was just not available in Korea. Domestic resources were limited to anthracite coal, some of which was being mined with a very fine particle size, which at the time could not be briquetted for use in iron production. By contrast, good coking coals were readily available on the world market and could be imported from the United States, Canada, and Australia. The U.S. coals were high in quality, and western Canada also had extensive reserves of good coking coals while the Australian supplies were in closer proximity. Japanese steelmakers had determined that excellent blast-furnace coke could be produced using blends of coal imported from these various sources. In like manner, POSCO has effectively blended imported coals in producing coke at both the Pohang and Kwangyang plants.

In contrast to the lack of domestic iron-ore and coking-coal reserves, limestone and dolomite resources are plentiful in Korea, and since the startup of Pohang's first blast furnace in 1973 have been relied upon to furnish large quantities of fluxing agents for ironmaking purposes. Over the last five years, POSCO has consumed an average of 7.0 million tonnes of domestic limestone each year.

IMPORTED RESOURCES

The near total dependency that a large-scale Korean integrated steel plant would have on imported iron ore and coking coal did not deter POSCO from building two such integrated plants. When the Pohang plant's construction started in 1970, neighboring Japanese steelmakers just across the East Sea were closing in on a production milestone of 100 million annual tonnes of crude steel. They had demonstrated most emphatically that a very large steel industry was able to thrive in a nation not having the requisite raw materials within its own borders.

POSCO's selection of the seacoast plant sites at Pohang and Kwangyang has served the company's raw-materials strategy by affording deep-water access to large bulk cargo carriers to drastically reduce the cost of importing the raw materials needed by the company's blast furnaces. As described in Chapter 2, POSCO operates large-scale wharves and raw-material storage yards at both plant locations for the receipt and stockpiling of the nearly 70 million tonnes of raw materials the company now purchases each year. The port facilities at Pohang and Kwangyang each can simultaneously accommodate multiple large vessels, including a maximum 250-thousand-DWT class carrier (See Figure 5.1).

During the last 20 years, POSCO has imported an aggregate 474.4 mil-

Figure 5.1 Unloading a Large-Bulk Ore Carrier

lion long dry tonnes (DLT) of iron ore, its yearly import volume having grown at a compound annual rate of 6.2 percent. As Table 5.1 shows, ore imports have risen from 11.6 million DLT in 1981, when the Pohang plant's fourth stage was completed, to 36.0 million DLT in 2000. Starting in 1987, when Kwangyang's first blast furnace came on line, import volume has trended higher, boosted by increasing ore requirements to feed each of the plant's four additional blast furnaces as they started production in 1988, 1990, 1992, and 2000. POSCO now feeds the 10 blast furnaces at its two integrated plants with high-grade iron ore from around the world and has also purchased an average of 413 thousand DLT of lower-grade domestic ore each year over the last five years.

POSCO's imported iron ore comes from a number of countries. Australia, the closest geographically, is also the leading source of supply, accounting for imports of 20.8 million DLT in 2000, representing 57 percent of the year's total imports of 36.0 million DLT. Brazil, the second-largest source, supplied 10.5 million DLT, or 29 percent of the total, followed by

Table 5.1 Iron-Ore Imports of POSCO: 1981–2000 (thousands of DLT)

Year	*Quantity*
1981	11,609
1982	11,454
1983	11,341
1984	12,410
1985	12,138
1986	12,656
1987	16,259
1988	18,710
1989	21,737
1990	22,532
1991	27,704
1992	28,858
1993	32,396
1994	31,481
1995	32,671
1996	32,585
1997	33,524
1998	33,534
1999	34,294
2000	36,013

Source: Pohang Iron and Steel Company, Ltd.; 1 DLT (long dry tonne) × 1.016047 = 1 tonne.
Please note that POSCO is not the sole importer of iron ore in Korea. Cement makers also import some iron ore. However, their portion of import is very small.

India with 2.6 million DLT, or a 7 percent share. The remaining 2.4 million DLT of ore came from a number of countries, chief among them China and Canada, which accounted for one million DLT and 600 thousand DLT, respectively.

After iron ore, coal is POSCO's next-largest raw-materials import, accounting for an average yearly import volume of 18.2 million tonnes over the last five years (See Table 5.2). As in the case of iron ore, Australia has been the leading source of imports, having supplied 9.8 million tonnes of POSCO's coal requirements in 2000, or 53 percent of the total. Notably, 2000 saw China displace Canada as the second-largest source of coal, the two nations supplying 3.8 million and 3.5 million tonnes, respectively. Coal imports from the United States, meanwhile, have undergone a five-year decline, from 2.7 million tonnes in 1996 to one million tonnes in 2000. The 18.6 million tonnes of total coal imports in 2000 included 669 thousand tonnes of anthracite coal.

In pursuing a raw-materials strategy aimed at securing stable sources of supply in international markets, POSCO has made significant investments to develop raw-material sources in other countries. To ensure stable supplies of coal, the company has acquired 20 percent interests in each of two overseas joint ventures. The first, Miller Pohang Company, was formed in 1982 to develop the Mount Thorley Coal Mine project in Australia, and POSCAN also was formed in 1993 to develop the Greenhills Coal Mine project in Canada. In 1996, a Brazilian joint venture, KOBRASCO, was formed to produce iron-ore pellets, with POSCO having a 50 percent ownership share. And finally, to secure chrome supplies essential to POSCO's growing stainless-steel business, a 25 percent share was

Table 5.2 Coal Imports of POSCO: 1996–2000 (millions of tonnes)

Country	*1996*	*1996*	*1998*	*1999*	*2000*
Australia	9.3	10.0	9.7	9.0	9.8
Canada	4.2	3.9	3.9	3.4	3.5
United States	2.7	2.6	2.3	1.7	1.0
China	1.0	0.8	1.3	2.6	3.8
Others	0.8	1.0	1.0	1.0	0.5
Total	18.0	18.3	18.2	17.7	18.6

Source: Pohang Iron and Steel Company, Ltd.

acquired in another 1996 joint venture, POSCHROME Pty. Ltd., of South Africa.

Over the last five years, POSCO has imported an average of 670 thousand tonnes of hot-briquetted iron or HBI each year. HBI is one of the primary charge materials for the electric furnace at the Kwangyang minimill. Most recently, POSCO has signed a letter of understanding with Australia's BHP to develop Mining Area "C" in western Australia as a new source of high-quality iron ore.

RAW MATERIALS TODAY

Having executed an effective raw-materials strategy since the early 1970s, POSCO is continuing to pursue measures that will improve the efficiency of its raw-materials purchases. The current strategy for doing so includes expanded yearly contracts and spot purchases and also the expanded use of more cost-effective materials, including those that can be acquired domestically and from the company's overseas joint ventures. To transport the raw materials purchased abroad, the company long has relied on vessels hired for its sole use on a long-term basis. In the early 1980s, to further stabilize transportation costs and to ensure on-time delivery, a "dedicated vessel" strategy was adopted whereby vessels are built and operated exclusively to ship the company's raw materials. Today, POSCO runs a fleet of 38 dedicated carriers with a combined deadweight capacity of some seven million tonnes to move about 80 percent of the raw materials it moves over water. Unless the oceans run dry, this will ensure that the company's steel plants continue to receive stable, cost-effective raw-materials supplies.

NOTES

1. Korea International Steel Associates (KISA), *Project Development Plan*, March 1967, pp. III–1f.
2. KISA, *op. cit.* p. III–7.

6

The International Strategy

POSCO has had no other choice than to pursue an international business strategy. This was ordained by Korea's near total lack of the basic raw materials needed by integrated steel plants, which demanded that the company do business internationally from day one, and which in 2000 required the company to make overseas purchases of iron ore, coking coal, scrap, and ferroalloys representing a combined volume of nearly 60 million tonnes. To help balance the international payments these raw-material purchases entailed, POSCO sold 6.4 million tonnes of steel products abroad in 2000, or 24 percent of the company's total steel sales, the other 76 percent going to support the continuing growth and development of Korea's steel-consuming industries.

POSCO's success in fulfilling its essential role as an exporter of steel has been made possible by its competitive costs and the quality of its products. However, the company's export success also has derived from a key international strategy, namely a broad-based participation in international joint ventures. Notably, overseas companies in which POSCO has ownership participation received 23 percent of all the steel products it exported last year. As both a buyer and seller in world markets, and also as one of the world's leading joint-venture participants, POSCO has spun a global network of commerce and has thereby created a truly global company.

STEEL-PRODUCT EXPORTS

During four of the last five years, POSCO's steel-product exports have, on average, accounted for about one-fourth of the company's total steel sales, the notable exception being 1998, when domestic demand slowed due to the Asian financial crisis, and the export share rose to the 30 percent level (See Table 6.1). During 2000, high levels of activity in domestic automaking, shipbuilding, and other primary steel-consuming industries boosted home-market sales by 5.5 percent to 20.4 million tonnes, causing a full percentage point shift in the ratio of domestic to international sales to the aforementioned 76 percent to 24 percent level. Over the 1996–2000 period, POSCO's total steel-product sales increased at a compound annual rate of 3.2 percent, with domestic and export sales increasing at rates of 3.7 percent and 1.7 percent, respectively.

The product breakdown of POSCO's export sales is shown in Table 6.2. Cold-rolled sheet products are the company's leading export, accounting for some 54 percent of total export sales in 2000. In second place, hot-rolled sheet products represented 31 percent of sales in the same year, but their export volume and share of the company's overseas business has been on the wane over the last five years, reflecting a strategy that has placed greater emphasis on the marketing of higher-value-added steel. The drive to boost value-added sales also has been reflected in the growth of electrical-steel and stainless-steel exports, which carry much higher price tags than many other steel products. Over the last five years, the company's overseas stainless business has more than tripled, and electrical-steel exports have undergone a more than four fold increase.

Starting in the early 1980s, POSCO implemented a strategy to diversify the geographic distribution of its overseas sales, which by 1985 had made Japan and Southeast Asia the company's leading export markets, a position formerly held by the United States. In more recent years, China has emerged as a growing outlet for the company's steel and now ranks second behind Japan among the foremost export markets (See Table 6.3). In 2000, nearly one-third of POSCO's steel-product exports were sold in

Table 6.1 POSCO's Domestic Sales vs. Exports: 1996–2000 (thousands of tonnes)

	1996	*1997*	*1998*	*1999*	*2000*
Domestic Sales	17,593	18,657	17,414	19,289	20,358
Exports	6,008	6,514	7,501	6,448	6,425
Total	23,601	25,171	24,915	25,737	26,783

Source: Pohang Iron and Steel Company, Ltd.

Table 6.2 POSCO's Export Sales by Product: 1996–2000 (thousands of tonnes)

	1996	*1997*	*1998*	*1999*	*2000*
Hot-Rolled Steel	2,510	2,388	2,627	2,203	1,977
Plates	299	304	424	356	353
Wire Rods	132	138	228	186	129
Cold-Rolled Steel	2,539	2,732	3,769	3,176	3,446
Electrical Steel	33	27	334	120	139
Stainless Steel	106	232	271	371	376
Others	389	693	148	36	5
Total	6,008	6,514	7,501	6,448	6,425

Source: Pohang Iron and Steel Company, Ltd.

Japan and 26 percent in China, with shares of 15 percent and 12 percent accounted for by Southeast Asia and the United States, respectively.

OVERSEAS JOINT VENTURES

POSCO currently engages in 23 steel-related joint ventures outside of Korea. They are located in 12 countries, and as Table 6.4 indicates, most of them were entered into in the decade of the 1990s. One of the company's principal joint ventures is USS-POSCO Industries, established in 1986 with USX, the largest integrated steelmaker in the United States, to process hot-rolled coils into high-quality sheet products, including cold-rolled sheets, galvanized sheets, and tinplate. USS-POSCO presently accounts for a majority of POSCO's U.S. steel-product sales. The ownership percentages held by POSCO in its 23 joint ventures vary considerably,

Table 6.3 POSCO's Export Sales by Region: 1996–2000 (thousands of tonnes)

	1996	*1997*	*1998*	*1999*	*2000*
Japan	2,451	2,508	2,142	2,135	2,089
China	1,008	1,057	1,459	1,549	1,677
Southeast Asia	1,401	1,518	1,180	917	936
United States	810	966	1,062	803	775
Others	338	465	1,658	1,044	948
Total	6,008	6,514	7,501	6,448	6,425

Source: Pohang Iron and Steel Company, Ltd.

Table 6.4 POSCO's Overseas Joint Ventures in Steel

Country	*Name of Joint Venture*	*Date Formed*	*% Share of POSCO Group*
China	Dalian POSCO-CFM Coated Steel Co., Ltd.	Nov. 1995	POSCO 40% POSTEEL 15%
	Zangjiangang Pohang Steel Co., Ltd.	Nov. 1996	POSCO 90%
	Zangjiangang Pohang Stainless Steel Co., Ltd.	Feb. 1997	POSCO 80%
	Sunde Pohang Coated Steel Co., Ltd.	Apr. 1997	POSCO 90%
	POS–Tianjin Coil Center Co., Ltd.	Oct. 1994	POSCO 10% POSTEEL 60%
	Guangzhou Jindo Container Mfg. Co., Ltd.	Jan. 1993	POSCO 10%
	Zhangjiang Xiaoshi Coil Co., Ltd.	Oct. 1996	POSTEEL 17.5%
	Suzhou Dongshin Color Sheet Co., Ltd.	Sep. 1995	POSTEEL 30%
	Shunde Xingpu Steel Center Co., Ltd.	Jul. 1997	POSCO 10.5% POSTEEL 10.5%
Vietnam	POSVINA Co., Ltd.	Apr. 1992	POSCO 50%
	VINAPIPE	May 1992	POSCO 10.6%
	VSC–POSCO Steel Corp.	Jan. 1994	POSCO 35% POSTEEL 5%
Thailand	POS–Thai Steel Service Center Co., Ltd.	Jan. 1997	POSCO 11.7% POSTEEL 52.2%
	The Siam United Steel Co., Ltd.	Jul. 1995	POSCO 3%
Japan	POSMETAL		POSTEEL 50%
India	POS–Hyundai		POSTEEL 19.5% POSCO 10%
Myanmar	Myanmar–POSCO Steel Co., Ltd.	Nov. 1997	POSCO 70%
Brazil	KOBRASCO	Mar. 1996	POSCO 50%
Venezuela	POSVEN	Apr. 1997	POSCO 40% POSTEEL 10% POSEC 10%
U.S.A.	USS–POSCO Industries	Apr. 1986	POSCO 50%
Canada	Fording Coal Ltd.	1993	POSTEEL 20%
Australia	Miller Pohang Co., Ltd.	1982	POSCO 20%
South Africa	POSCHROME (Pty) Ltd.	Jul. 1996	POSCO 25%

Source: Pohang Iron and Steel Company, Ltd.; includes joint ventures engaged in steel manufacturing and in the production of raw materials for ironmaking and steelmaking.

ranging from a high of 90 percent in two Chinese companies, Zangjiangang Pohang Coated Steel and Sunde Pohang Coated Steel, to a low of 3 percent in a joint venture in Thailand, namely the Siam United Steel Company, Ltd.

Significant tonnages of steel, principally sheets, are sent by POSCO to 18 of its joint ventures that are engaged in steel-product manufacturing. In 1997, the aggregate amount shipped to these companies was 996 thousand tonnes. In 1998, the figure was 1.16 million tonnes; in 1999, it was 1.39 million tonnes; and in 2000, it was 1.46 million tonnes. This indicates the fairly large extent to which POSCO participates in processing steel in many locations around the world. One of the primary locations is China, where POSCO has nine joint ventures to which it shipped 480 thousand tonnes of various sheet products during 2000.

In addition to participating in joint ventures in the steel business, POSCO, as discussed in Chapter 5, has ownership shares in several raw-materials joint ventures. They are five in number and include a coal-mining company in Australia in which POSCO has a 20 percent interest and an iron-ore pellet plant in Brazil in which POSCO's ownership interest is 50 percent. The Brazilian joint venture, KOBRASCO, produces four million annual tonnes of iron-ore pellets of which POSCO is allocated 2.3 million tonnes for use mainly by its No. 5 blast furnace at Kwangyang, which provides molten iron for electric-furnace steelmaking at the plant's thin-slab minimill. HBI from a Venezuelan joint venture, POSVEN, also is supplied to the Kwangyang minimill and to the basic-oxygen steelmaking shops at both Pohang and Kwangyang.

Joint ventures have been a key part of POSCO's international strategy, not only assuring steady supplies of essential raw materials, but also contributing to the company's export business and affording opportunities to forge close cooperative relationships with overseas steel companies in order to better serve their common customers.

7

The Profit Strategy

It is fitting that POSCO's profit strategy be considered in one of this book's concluding chapters. The ability to consistently post earnings in the profit-deprived steel business, which is exactly what the company has done, represents conclusive confirmation, bottom-line proof if you will, that all of the other strategies already discussed here have been well chosen and are extremely effective. Conversely, POSCO's profit strategy has been to ensure and coordinate the effective execution of its other strategies, from those employed to build and operate the company's two integrated steel plants, to those guiding its customer relationships and the sale of its products, its research and development programs, the welfare and education of its employees, the acquisition of its raw materials, the conduct of its export business and its international joint ventures, and finally, its long-term quest to attain privately owned status.

POSCO's profit strategy represents an affirmation of a core principle behind the company's business philosophy and the development of its corporate culture from the days of its founding some 33 years ago. This core principle was codified upon the company's incorporation under Korean commercial law as a profit-based entity, which represented a decided departure from tradition for what at the time was a government-owned firm. Today, after more than three decades in business, POSCO remains true to its charter, ever striving to operate efficiently and profitably, however difficult the world steel environment becomes. The company's singular success in regularly achieving this goal has made possible a complete transition from government ownership, and today the company stands as a 100 percent privately owned business enterprise.

THE PROFIT RECORD

POSCO's profit record has been most extraordinary (See Table 7.1). Year 2000 net earnings set a new all-time high in terms of Korean won, marking the 28th consecutive year in which the company reported profitable results. The last time a loss was posted was in 1972, when the company's first integrated plant at Pohang was still undergoing the first stage of its construction. Not until July 1973 was the first stage completed, but remarkably, the company was able to report a profit for the full year, something which has been a recurring event every year since.

During 1988–89, when conditions improved in the world-steel market and the second-stage construction of the Kwangyang plant was completed in July, 1988, POSCO's profits broke the $100 million mark for the first time. During the last three years, despite difficult conditions faced by steelmakers in many countries around the world, POSCO's profits approached and then surpassed the billion-dollar level, with won-denominated net profits for 2000 reaching a record 1,637 billion won. This excellent performance was achieved despite conditions of oversupply in the world steel market generally, intensified competition, and serious volatility in world energy prices.

The net earnings of 1,637 billion won posted for 2000 were up 5.1 percent from 1,558 billion won in 1999 and represented a 14 percent return on sales of 11,692 billion won. Sales were up 9.3 percent from the 1999 level, lifted by a 4.1 percent gain in the tonnage volume of steel products sold to 26.8 million tonnes and an average improvement in selling prices of 21 thousand won per tonne. Although the cost of goods sold was up 7.9 percent over 1999 levels, it was more than outpaced by gross profit, which was up a strong 14.4 percent to 2,637 billion won. Operating income was up a substantial 15.4 percent to 2,099 billion won, and operating margins edged up from 17 percent to 18 percent.

POSCO attributed these improvements behind its 2000 record profitability to four major factors: 1) a decrease in the cost of goods sold due to lower depreciation expenses, 2) increased sales of higher-margined steel

Table 7.1 POSCO's Profit History: 1968–2000

	Korean won	foreign exchange rate Korean won/U.S. dollars	U.S. Dollars
1968	---	---	---
1969	---	---	---
1970	---	---	---
1971	---	---	---
1972	−550,770,793	395.67	−1,391,995
1973	4,639,391,575	399.31	11,618,521
1974	35,464,514,584	413.19	85,831,009
1975	9,128,071,300	484.19	18,852,251
1976	15,299,221,367	484.84	31,555,196
1977	13,721,953,307	485.00	28,292,687
1978	21,894,602,392	484.97	45,146,303
1979	26,219,941,557	484.95	54,067,309
1980	10,808,893,943	659.90	16,379,594
1981	44,292,705,397	700.50	63,230,129
1982	15,302,469,985	748.80	20,435,991
1983	52,074,344,806	795.50	65,461,150
1984	60,191,714,279	827.40	72,748,023
1985	61,729,047,684	890.20	69,342,898
1986	62,010,432,641	861.40	71,987,965
1987	70,331,397,770	792.30	88,768,645
1988	134,356,635,806	684.10	196,399,117
1989	144,511,352,563	679.60	212,641,778
1990	79,025,204,482	716.40	110,308,772
1991	145,680,235,649	760.80	191,482,960
1992	185,061,312,793	788.40	234,730,229
1993	294,617,035,811	808.10	364,l579,923
1994	383,221,067,473	788.70	485,889,524
1995	839,707,294,232	774.70	1,083,912,804
1996	623,961,686,620	844.20	739,115,952
1997	728,983,024,792	1415.20	515,109,543
1998	1,122,867,118,025	1207.80	929,679,680
1999	1,558,031,776,383	1145.40	1,360,251,245
2000	1,637,991,067,182	1252.50	1,306,978,896

Source: Pohang Iron and Steel Company, Ltd.; data indicate nonconsolidated net profits; exchange rates are those occurring at year end.

products such as stainless steel, 3) a further reduction in production defects and product rejection rates, and 4) the company's continuing cost-reduction efforts. The company concluded that the "continued upward trend in financial performance is a direct result of our ongoing efforts to improve both our profitability and financial structure through innovations in business efficiency."[1] Both POSCO's long-term earnings performance and the strong results posted in the last three years have been instrumental in attracting private investors to enable the company to complete its quest toward full privatization.

PRIVATIZATION

When POSCO was incorporated on April 1, 1968, under Article 288 of Korea's Commercial Code, the company was organized as a private, profit-based enterprise, wherein all investment and operating decisions were under the control of management. At the same time, the special legislation required to provide the company with essential government funding placed majority ownership under government control. Over the more than three decades that have now passed, POSCO has been on a long journey toward privatization, working to establish itself as a company with a proven record of profitability that would be attractive to new investors.

One of the first major breakthroughs toward privatization came in June 1988, when 36.4 percent of the company's shares owned by the Korean government and the state-run Korea Development Bank were sold to the public, whereupon POSCO was listed on the Korea Stock Exchange. As a result of this transaction, the ownership interest in the company controlled by the government and the Korea Development Bank declined to 35 percent from 71.4 percent. Another major step toward privatization occurred in October 1994, when POSCO American Depository Receipts (ADRs) were listed on the New York Stock Exchange, and one year later, the ADRs were listed on the London Stock Exchange.

Further advancing POSCO's privatization, the Korean government in July 1998 announced a plan to completely privatize 11 public corporations, including POSCO, by selling the government's remaining equity in each of them. Five of the corporations, including POSCO, were to be completely privatized in one stage, while the other six were to be sold off gradually. In POSCO's case, all the remaining ownership interest of the government and the Korea Development Bank were to be sold to both domestic and foreign investors, with individual purchases limited to a 3 percent ownership share to prevent any single investor from obtaining undue influence over the company. Once complete privatization was accomplished, this 3 percent limit was to be abolished for foreign investors, and it was set to expire for all investors at the end of 2001.

POSCO shares held by the government and the Korea Development Bank were issued as ADRs and sold on the New York Stock Exchange in December 1998 and in July 1999. Following these transactions, the Korea Development Bank announced that its remaining 12.84 percent interest in POSCO would be sold at auction to Korean investors. Shares representing this interest were eventually sold to POSCO and to foreign investors between December 1999 and October 2000, when the final 6.84 percent share was purchased, bringing to an end the long journey toward privatization. On October 4, 2000, POSCO was reborn as a totally private company.

NOTE

1. Pohang Iron and Steel Company, Ltd.; *Annual Report*, 2000, p. 26.

8

The Future Strategy

As POSCO moves into the 21st century as a totally private enterprise standing at the apex of the world steel industry, the company is not, in its own words, "content to just sit back and enjoy the status quo." Instead, a future strategy has been adopted with a mid-range goal to more than double the company's value over the next five years. This strategy has three major components, namely, to complete implementation of a new, state-of-the-art integrated information system, to secure and build upon the company's worldwide competitive edge in steelmaking, and to expand the company's business interests in several promising new industries.[1]

Under the first part of the future strategy, POSCO has worked on the completion of a new process-innovation or PI initiative, which involved two and a half years of companywide effort and has at its core a new digital-management system called POSPIA. On July 2, 2001, the new system went live on a big-bang basis, its name, POSPIA, representing a combination of POSCO and Utopia, to signify that the PI initiative has provided the company with the ideal in business-management systems.

POSCO started the PI initiative by analyzing "As-Is" to identify the company's strengths and weaknesses, and by designing "To-Be," the ideal to be achieved. Such master planning was fundamental to building an ERP/SCP (Enterprise Resources Planning/Supply Chain Planning) system, which was placed through nine months of integration testing prior to launch. In this respect, POSPIA integrates POSCO's e-business, both e-sales and e-purchasing, across the company's value chain over the Web. This creates a win-win situation for POSCO, its customers, and suppliers, thereby empowering the company in its effort to maintain a position of global leadership.

PI and POSPIA together represent a true business innovation that departs from past product-centered practices and fully restructures the company's information-technology system, business processes, corporate organization, and employee focus around service to the customer. In so doing, lead times for sales planning will dramatically be reduced from 60 to 15 days, which means that customers can determine product availability from POSCO a month and a half earlier than before. In addition, the lead time for hot-rolled steel from order entry to product delivery will be shortened from 30 to 14 days, allowing customers to secure their steel faster and thereby reduce their average time of inventory possession.

POSPIA also provides true "one-stop service" by enabling the customer to view information and process transactions in real time over the Web. The system's "one-stop service" screen presents 98 types of information, ranging from the status of individual customer orders to POSCO's plans for production and sales, permitting customers to place and track their orders on-line. Further, the system is designed to enhance relationships with the company's suppliers, assist decision making, and promote organizational efficiency, all of which are expected to add to the company's earnings and value.

The second component of POSCO's future strategy, which is aimed at solidifying the company's competitive edge in steel, is backed up by an expanded budget for facilities investment of 1,600 billion won in 2001. A major portion of this budget is directed at upgrading the ratio of value-added steels in the company's product mix. As detailed below, the company is expanding its stainless- and electrical-steel capacities as part of the value initiative, but it is also adding to its hot-rolled-strip capacity in an effort to retake a domestic market segment recently captured by imports. In addition, the company will continue to invest in environment-friendly, energy-saving technologies, and will spend 1.9 percent of its sales on research and development to construct its FINEX ironmaking demonstration plant.

The third component of the future strategy aims to reinforce POSCO's solid foundation for future growth and the creation of corporate value. One approach to advancing this goal will be to accelerate joint research and development projects the company has underway in its August 2001 strategic alliance with Nippon Steel. And the company will continue to expand its interests in such new businesses as multimedia communications, energy, bioscience, and other strategic fields.

VALUE IN STEEL

In January 2001, the chairman and CEO of POSCO, Sang-Boo Yoo, held a press conference at which he announced plans for the company's future. Part of the future strategy set forth in his announcement consists of a significant investment program covering several projects designed to improve existing facilities, add capacity, and most importantly, increase the company's ability to produce value-added steel. In all, some six projects were enumerated, covering facilities at both the Pohang and Kwangyang plants (See Table 8.1). The total investment cost of these projects amounts to 1,145.4 billion won or $914.5 million at the year-end 2000 rate of exchange.

Part of POSCO's investment plan is for energy reclamation by means of installing two coke-dry-quenching (CDQ) units, one at each of the company's plants. Work on the CDQ unit at the Pohang was started in November 2000 and is scheduled for completion in April 2002, while at Kwangyang, construction started in May 2001 and will be completed in October 2002. The energy reclaimed by installing the units will include 53 thousand MWH/yr of electricity at Pohang and 130 thousand MWH/yr at Kwangyang; in addition to which the units will permit the recovery of considerable quantities of low-pressure steam for subsequent use at both plants.

POSCO intends to increase its production of hot-rolled stainless-steel sheets by 450 thousand tonnes per year. This will raise Korea's production, which was 1.7 million tonnes in 1998 to nearly 2.2 million tonnes. Korea has been producing stainless steel in significant tonnages since 1990. In that year, it produced 356 thousand tonnes, which was virtually doubled in 1991 to 629 thousand tonnes. Stainless-steel output reached the million-tonne mark in 1994, when it stood at 1.2 million tonnes. To satisfy the growing demand for stainless steel, POSCO's additional capacity is scheduled for completion by May 2003. Much of the construc-

Table 8.1 Major Investment Plan for POSCO

	Project	*Expected Effects*	*Construction Period*
Energy	#4-2 CDQ Installation of Pohang Works	—Electricity Reclamation: Pohang 53, Kwanyang 130 thousand MWH/year	'00.11–'02.04
	#2 CDQ Installation of Kwangyang Works	—Low-Pressure Steam Reclamation: Pohang 405, Kwangyang 494 thousand tons/year	'01.05–'02.10
Enhancement of Production Capacity and Quality	Increase of STS HR Production	—Capacity Increase of STS HR by 450,000 tpy	'01.07–'03.05
	Rationalization of Silicon Steel Mill of Pohang Works	—Capacity Increase of GO (grain oriented) by 25,000 tpy from 35,000 to 60,000 tpy —Capacity Increase of NO (non-oriented) by 60,000 tpy from 470,000 to 530	'00.04–'01.11
	Rationalization of #2 Hot-Strip Mill of Kwangyang Works	—Capacity Increase of HR by 470,000 tpy from 3.6 million to 4.1 million tpy	'00.09–'02.07
Development of New Technology	Construction of FINEX Demo Plant	—Capacity: HB1 600,000 tpy —Application of new environmentally friendly process	'00.12–'03.02

Source: Pohang Iron and Steel Company, Ltd.

tion and the capital expenditures for the stainless project will be concentrated in 2002.

Stainless steel is used in a number of industries where resistance to corrosion is required. Among these industries are automotive, construction, appliances, and air conditioning. In the construction of multistory buildings, stainless is used in elevators and some special types of doors, as well as in the façade. It is also used extensively for roofing. A considerable tonnage of stainless is shipped to processing industries. These shipments are, to a great extent, in the form of hot-rolled sheets, which can be processed into cold-rolled and colored sheets. Processing companies also modify the sheets by slitting and cutting to length. The automotive industry uses stainless for exhaust pipes and also for steel used in automobile frames. A large tonnage of stainless is used for food processing, as well as for cutlery. Hot-rolled stainless has a minimum use as such and usually undergoes further processing by someone other than the steelmaker. Service centers buy a large tonnage of stainless and distribute it to a great number of industries in relatively small amounts.

Another of POSCO's ongoing investment projects is designed to increase grain-oriented varieties. Grain-oriented steel, which is the stronger of the two types, has approximately 5.0 percent silicon in its composition. POSCO's current capacity for grain-oriented production is to be increased by 25 thousand tonnes annually to 60 thousand tonnes. The second type of silicon steel, non–grain oriented, contains less silicon, its content varying from 1.5 to 3.0 percent.

Grain-oriented steel is a premium-grade silicon-alloy product in which the inherent magnetic properties of the metal are developed to the maximum by rearranging the granular structure of the steel in a specified pattern. Steel has a grain structure which can be observed only under a specially designed microscope. The grain structure in steel that has received no special heat treatment will appear to be an extremely random one, lacking any particular pattern. In some steel, this patternless arrangement of the grains is of little or no consequence to the ultimate use of the product, but it has been demonstrated that in the case of silicon steel, magnetic properties, which are of primary importance, are more fully developed when the grain structure is modified to align the grains in a particular way.

Consumers of grain-oriented steel are able to take advantage of its directional grain alignment by constructing their electrical equipment in such a manner that the flow of the magnetic flux is in the rolling direction of the sheets incorporated into the laminated cores of the equipment. One significant effect is to measurably reduce core losses and to increase steel permeability. In transformer manufacture, for example, this makes it possible to reduce unit size without sacrificing electrical efficiency and with direct savings in copper laminations and steel weight.

In the production of grain-oriented steel, exacting care must be exercised in every phase of the production process, but particularly in the annealing phase. All silicon steel is heat treated or annealed after reduction to the desired thickness. Annealing has several desirable effects, such as the recrystallization and growth of grains in the steel, relief of rolling stresses, and he reduction of carbon content. Although such results are important in the manufacture of all silicon grades, grain-oriented steel calls for more extensive and critical annealing. For example, where some silicon grades may be produced with just one hot or cold rolling and one annealing, the production of grain-oriented steel involves a two-stage cold reduction and two or more annealings in furnaces where both heat and atmosphere are rigidly controlled. It is in these latter units that grain-orientation takes place.

Some of the savings attributed to grain-oriented steel in transformer manufacturing are considerable, such as 25 percent of core steel, 25 percent of copper, and 20 percent of structural and tank steel. These savings are of considerable significance, particularly since one of them is copper. Thus, grain-oriented steel is much more expensive to produce than the non-grain-oriented type and sells for a much higher price, even though smaller amounts are sold.

POSCO's investment program for the future includes a project to increase the capacity of the No. 2 hot-strip mill at the Kwangyang plant by 470 thousand annual tonnes, from a current level of 3.6 million to 4.1 million tonnes. The hot-strip-mill modifications at Kwangyang are scheduled for completion in July 2002.

POSCO's investment plan will permit the company to expand its participation in a number of steel markets, such as automotive and appliances, both in Korea and abroad. POSCO sold 26.8 million tonnes of steel products in 2000, and the new program will increase its product output by an aggregate of two million tonnes. Further, the equipment will be the newest available and, thus, will make it possible for POSCO to continue to enhance the quality of its products, while also serving the company's ongoing strategy of upgrading its product mix to include a higher proportion of value-added steels.

A BRIGHT FUTURE

Having been the driving force behind Korea's industrialization over the past three decades, POSCO now looks ahead to the promises and challenges of a new century as the world's number one steel-producing company. In achieving this status, the company is unique in the annals of steel history, having risen to the top solely through its own internal growth and development. Ever since POSCO was founded, and from the first day in April of 1970, when a shovel of earth was first turned near a small fishing village called Pohang, the company has prospered through visionary leadership and the dedication, hard work, and sacrifice of its 19,928 employees. The same unique spirit that has built the company is reason to have confidence that POSCO will succeed in shaping a bright future for itself and the nation it serves.

NOTE

1. Pohang Iron and Steel Company, Ltd. *Annual Report* 2000, pp. 3 and 7.

Bibliography

American Metal Market. New York: American Metal Market LLC, various daily issues.
Business Korea, November 1998.
Cooke, Henry, ed. *Iron and Steel Works of the World,* 13th edition. Surrey, England: Metal Bulletin Books Ltd., 1999.
Hanil Can Company. *Welcome to Hanil Can.* Seoul: Hanil Can Company, 1998.
Hogan, William T. "A Truly Global Company." *New Steel* (May 1997): 82–95.
———. *Capital Investment in Steel: A World Plan for the 1990s.* New York: Lexington Books, 1992.
———. "Future Locations for Asian Steel." *Steel Technology Asia 1998.* London: Sterling Publications Limited, 1998.
———. *Global Steel in the 1990s: Growth or Decline.* Lexington, Massachusetts: Lexington Books, 1991.
———. "Joint Venture Developments in Asia." *Asia Steel, 1997.* London: Crambeth Allen Publishing, 1997.
———. "Pohang Steel Co. Continues to Grow," *Iron and Steel Engineer* (April 1985): 32–38.
———. *Steel in the 21st Century: Competition Forges a New World Order.* New York: Lexington Books, 1994.
———. "South Korean Steel Growth Spurs Industrial Development." *Iron and Steel Engineer* (November 1990): 29–34.
———. "The Challenge to Steel from Substitute Materials." *Steel Technology International 1993,* London: Sterlng Publications Limited, 1993.
Huskonen, Wallace D. "POSCO Goes 'Mini' to Grow Larger." *Metalproducing* 33 (July 1998): 30–36.
Innace, Joseph J. and Dress, Abby. *Igniting Steel: Korea's POSCO Lights the Way.* Huntington, New York: Global Village Press, 1992.
International Iron and Steel Institute, Committee on Statistics. *Steel Statistical Yearbook.* Brussels: IISI, various years.
Kim, H. Edward, ed. *A Miracle on Yongil Bay.* Seoul: Pohang Iron and Steel Company, Ltd., 1983.

Korea International Steel Associates. *Project Development Plan*. Unpublished Study, March 1967.

Korea Iron and Steel Association. *Bulletin Monthly*. Seoul: KOSA, various monthly issues.

———. *The Steel Industry in Korea*. Seoul: KOSA, various annual issues.

Metal Bulletin. London: Metal Bulletin Journals Ltd., various issues.

Pohang Iron and Steel Company, Ltd. *Annual Report,* various years.

———. *Dedication of Kwangyang Steel Works*. Seoul: Pohang Iron and Steel Company, Ltd. May, 1987.

———. *Fact Book: POSCO 2001*. Seoul: Pohang Iron and Steel Company, Ltd., 2001.

———. *The Necessity of the Construction of Kwangyang Works from a National Economic Point of View*. Unpublished company analysis, March 1983.

POSCO Bulletin. Seoul: Pohang Iron and Steel Company, Ltd., various issues.

POSCO Newsletter. Seoul: Pohang Iron and Steel Company, Ltd., various issues.

Seo, K. K. *The Steel King*. New York: Simon & Schuster, 1997.

Skillings, David N., Jr. "POSCO Operating Modern Steelmaking Complex in Pohang and Expanding with World's Largest Greenfield Steel Project at Kwangyang." *Skillings Mining Review* (January 11, 1986): 14–22.

The Korea Herald, May 26, 1999, and April 10, 2000.

U.S. Department of Energy, Energy Information Administration. *Country Analysis Briefs: South Korea*. Washington, D.C.: U.S. Government Printing Office, September 2000.

Ward's Automotive International. Southfield, Michigan: Ward's Communications, various issues.

Ward's Automotive Reports. Southfield, Michigan: Ward's Communications, various issues.

Zschetzsche, Albert; Schenk, Johannes L.; Eberle, Anton F.; Lee, Jun H.; Jung, Sun K.; and Lee, Il Ock. "The FINEX Fine-Ore Smelting Reduction Process." *Millenium Steel 2000*. London: Millenium Steel Publishing, 2000.

Index

Page numbers in italic refer to tables.

About the Author

Father Hogan graduated from Fordham College cum laude and received his M.A. and Ph.D. in economics from Fordham University, where he is currently a professor of economics. He has engaged in economic studies the steel industry and other basic heavy industries for over 40 years. He has authored a number of books, including *Productivity in the Blast Furnace and Open-Hearth Segments of the Steel Industry,* the first detailed study on the subject of steel productivity; *The Development of Heavy Industry in the Twentieth Century;* and *Depreciation Policies and Resultant Problems.* Over the years, he has also had more than 100 articles published, as well as a total of nineteen books.

In 1950, Father Hogan inaugurated the Industrial Economics Research Institute at Fordham University, of which he is the director. The Institute has produced studies for such agencies as the United Nations, World Bank, and a number of branches of the U.S. government, as well as for steel companies dealing with industrial economic problems. During the past 30 years, Father Hogan has visited most of the steel-producing facilities in the world and has delivered papers at steel conferences worldwide. In 1990, the Association of Iron and Steel Engineers established the William T. Hogan Lecture Series in recognition of his contributions to the steel industry.

He has been honored with the Gary Memorial Medal from the American Iron and Steel Institute; a Distinguished Life Membership from the American Society of Metals International; the Gold Tower of the Order of Industrial Merit award from the Republic of Korea; and honorary memberships from the Association of Iron and Steel Engineers, the International Iron and Steel Institute, and the Korean Iron and Steel Association. In 1996, Father Hogan was awarded the honorary degree of Doctor of Laws by Fordham University.

www.ingramcontent.com/pod-product-compliance
Lightning Source LLC
LaVergne TN
LVHW092132260926
848997LV00001B/2
9780739103012